AIR ORGANIZATIONS OF THE THIRD REICH:

The Luftwaffe

BY ROGER JAMES BENDER

ASSISTED BY HUGH PAGE TAYLOR

Schiffer Military History
Atglen, PA

Library of Congress Catalog Number: 92-83792

Printed in the United States of America.
ISBN: 0-88740-474-X

We are interested in hearing from authors with book ideas on related topics.

This book was originally published by
R. James Bender Publishing, 1972.

Published by Schiffer Publishing Ltd.
77 Lower Valley Road
Atglen, PA 19310
Please write for a free catalog.
This book may be purchased from the publisher.
Please include $2.95 postage.
Try your bookstore first.

Table of Contents

Introduction

"Air Organizations of the Third Reich" was published in 1967, at which time I felt that it was a significant effort in the research of uniforms and insignia. During the next few years, however, I realized that I had only touched on the subject. Due to determined research and by the ample generosities afforded me by fellow enthusiasts, the extensiveness of the material culminated. This study of the Luftwaffe reflects those years of discovery.

"Air Organizations", volume I, pertained to several related subjects and was contained in 192 pages. This new work on the Luftwaffe, due to the abundance of information, was only reluctantly held to 320 pages. This expanse holds true for all of the organizations originally covered in volume I. Rather than continue the numbering sequence of the series, I decided that the title should be altered to make it more descriptive of subject matter within (i.e. "Air Organizations of the Third Reich: THE LUFTWAFFE"). Future volumes of the "Air Organizations" series will encompass subjects as: The Foreign Legions, Legion Condor, N.S.F.K., etc.

The reader should note that units have not been given detailed historic or order-of-battle coverage (i.e. Regiment "General Göring", Parachute Troops, etc.) for these would be books in themselves. Therefore, segments of historical background were covered only to show the circumstances, dates and events which had a direct relationship with insignia and uniform introduction and evolution.

Roger James Bender
May, 1972

Acknowledgements

Never before has it been necessary to thank certain individuals, whose personal collections, reference libraries and photos were promptly made available, for their cooperation and active assistance. Among the major contributors were: Hugh Page Taylor, a good friend and invaluable colleague in this work; Friedhelm Ollenschläger, the man who transformed a maze of documented facts and dates into readily understood information; Warren Odegard, who graciously loaned me his invaluable "Luftwaffe-Verordnungsblätter", which you will see reference to on almost every page of this book; and John R. Angolia, who immediately sent albums of photos in response to my specific needs.

I also want to thank the following individuals who contributed not only time, energy and knowledge, but encouragement and guidance during the development of this book.

Philip H. Buss, B.A (Hons.)
Dick Deeter
Ronald Kwan
Richard Mundhenk
George A. Peterson
Helmut Schilling
Guenther Schoen
John Shea
Andy Southard
Robert Strodel
Garth Thompson
Vincent Wagner
Jerry Weiblen
Baron Theobald von Zastrow

NOTE:

The cover photo (pennant in the Robert Strodel Collection) was taken by Andy Southard, a good friend and fine photographer.

Author's Request

The use of photograph's and documentation from the many contributors to this and previous books has proved invaluable. But the never ending search for new material continues. Needed are photographs depicting all branches of the Wehrmacht and Party organizations (personalities, equipment, events, action, etc.). I am searching for individual photos or entire collections (including motion pictures) which I could either purchase or accept on loan (copies will be made and the originals returned within two weeks). Full credit will be given for photos used in future publications. Thank you for your consideration.

Roger James Bender
PO Box 23456
San Jose, CA 95153

In 1935, Luftsportgruppe 14 from Munich, organized the very successful "Alpenflug" with a showing of 75 sports aircraft.

Deutscher Luftsport-Verband

THE DLV UNIFORM

In March 1933, the Deutscher Luftfahrt-Verband incorporated all civilian flying clubs into one organization titled the Deutscher Luftsport-Verband, with Bruno Loerzer as its commander.[1] This new organization aggresively pursued its task of promoting civilian interest in model, glider, balloon and motored aircraft flight. They also educated the youth and schooled them in aviation fundamentals. The DLV was initially divided into 16 Landesgruppen (Including the free city of Danzig). In September 1933, The DLV also incorporated the former SA- and SS "Fliegerstürme" which had been formed earlier that year in July.

The commander of SS-Fliegerstaffl Süd greeted by Hitler at the Munich airport (1933). Note SS/SA pilots' wing over the right breast pocket.

SS/SA pilots' wing

An order dated November 4, 1933, decribes the DLV uniform. Its color was grey-blue and as Walking-Out Dress (Strassenanzug) consisted of: tunic, peaked cap, long trousers, blue shirt, black tie, grey gloves and belt with cross-over strap. While on duty, a Working-Dress could be worn which was as above but with breeches, riding boots and a forage cap.

Members of the Luftwaffe's secret officer corps (from 1.9.1933 - 28.2.1935) adopted the uniform of the DLV. A short time later other ranks were permitted the uniform and rank title of the Luftsport-Verband. The only exception was that shulder straps were worn on both shoulders, whereas the DLV wore a shoulder strap only on the right shoulder. Officers wore a brown "Sam Browne" belt (belt with cross-over strap) with double-claw buckle and other ranks wore a black "Sam Browne" belt with rectangular buckle.

[1] Loerzer was, in World War I, commander of Jagdgeschwader 3 and holder of the Pour le Mérite (the Blue Max).

DLV Service Uniform

The insignia at right was worn by all ranks on the upper left arm of the tunic and great-coat. The insignia was in gold wire for the top two ranks (grades 1 & 2), silver for other officers (grades 3-8) and aluminum-colored thread for all other ranks (grades 9-16).

DLV Formal Full-Dress Uniform

Unattributed early Fliers' insignia for motored aircraft.

NOTE:

Informal Full-Dress (Kleiner Gesellschaftsanzug) consisted of peaked cap, service tunic, trousers, white shirt, black tie, white gloves and ribbon bar. Formal Full-Dress for evening (Grosser Gesellschaftsanzug) was as above but for grades 9-16, the belt without the cross-over strap was worn. For grades 1-8, a brocade belt, aiguillette and medal bar were worn.

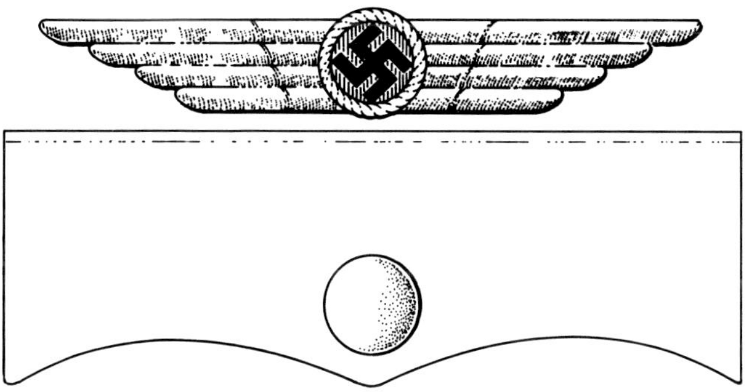

Pilot-Observer (Flugzeugführer-u. Orterabzeichen) Wing in silver embroidery

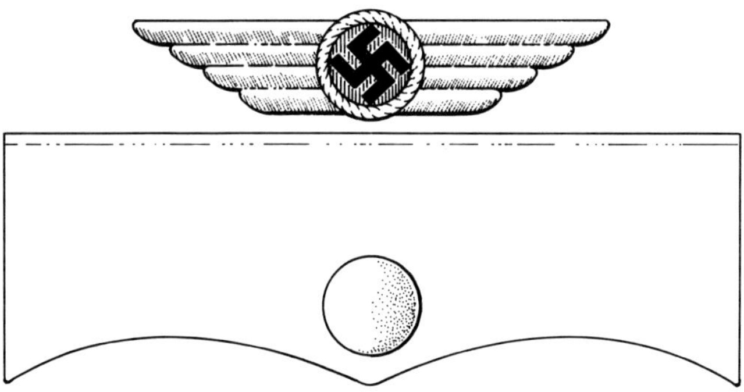

Wireless operator (Bordfunker) wing in silver embroidery

A set of Pilot/Observer or Wirless Operator wings were worn above the right breast pocket. These wings have also been observed worn over the left breast pocket but no regulations have been discovered to give credence to this deviation in dress.

The collar piping and uniform buttons for grades 1 & 2 were in gold; for grades 3-8 in silver; and for grades 9-16, the collar piping was in appropriate Waffenfarbe and the uniform buttons were in silver. The Waffenfarbe listing is as follows:

(a) Reichsminister für Luftfahrt
Staatssekretär für Luftfahrt } white

(b) Reichsluftfahrtministerium (Aviation Ministry Staff) – black

(c) Deutsche Verkehrsfliegerschule (German Commercial Flying School) . . . this section of the DLV was in fact the clandestine Luftwaffe in secret training – yellow

(d) All other sections of the Deutscher Luftsportverband – light blue

Flying Personnel Badge
(Fliegendes Personal)

Balloonist Badge
(Ballonfahrer)

Note the Balloonist Badge worn on the left sleeve of the DLV-Flieger in foreground.

Technical Personnel Badge
(Flugzeug Personal)

NOTE:

The above specialty badges were worn on the lower left sleeve of the tunic and greatcoat. They were authorized for non-commissioned ranks only (grades 9-16) and were in an aluminum colored thread on grey-blue.

Enlisted man's belt buckle

Officer's belt buckle as worn with brocade belt (DLV formal full-dress uniform)

REICHS-LUFT-AUFSICHT

The Reichs-Luft-Aufsicht, which was formed in 1935 by Göring, was responsible for proper flying operations with personnel on every airfield. While on duty, a Luft-Aufsicht observer would watch take-offs and landings and make notes on individual flying durations.

Their uniform was as that of the Deutscher Luftsport-Verband (later the Luftwaffe) but with a light green Waffenfarbe, and was discontinued later in the war. A special Reichs-Luft-Aufsicht gorget was worn for the duration of an observer's daily duties. Shoulder straps were worn on both shoulders. (see photo at right).

This photo of the Flugmeister was taken in April, 1935.

Reichs-Luft-Ausficht Gorget

DLV HEADGEAR AND INSIGNIA

DLV Cap Eagle

Aluminum peaked cap insignia (gold for grades 1 & 2) with red, white and black cockade.

First pattern silver peaked cap insignia with a black swastika and a red circle. This insignia was also utilized by "Abteilung Segelflug."

The basic DLV peaked cap was blue-grey with a black cap-band. The two top ranks (grades 1 & 2) had gold insignia, gold buttons and gold cap cords; grades 3-8 had silver insignia, silver buttons and silver cap cords; and grades 9-16 had silver insignia, black buttons and a black leather cap strap. The piping for all ranks was the same color as the cap. Members of "Abteilung Sefelflug" wore a peaked cap which was blue-grey overall. The DLV also had a blue-grey forage cap with the DLV eagle affixed on its front.

DLV glider personnel campaigning for members during Luftsport Recruitment Week in 1934. Note working-dress uniform.

"ABTEILUNG SEGELFLUG" UNIFORM

Working-Dress uniform

(1)

(2)

(3)

(4)

(5)

"ABTEILUNG SEGELFLUG" INSIGNIA

The short-lived section of the Deutscher Luftsport-Verband existed from late 1933 to 1935. It had a unique ranking system and insignia which was worn with the Working-Dress. A variation of this rank insignia was worn above the right breast pocket when the Working-Dress was shirt only and no jacket (see below).

The "Segelflug" ranks were as follows:

(1) Scharführer
(2) Truppführer
(3) Sturmführer
(4) Referent bei Landesverband (embroidered in silver)
(5) Leiter der "Abteilung Segelflug" beim Präsidium

***Collar patches have blue base**

"Abteilung Segelflug" Working-Dress

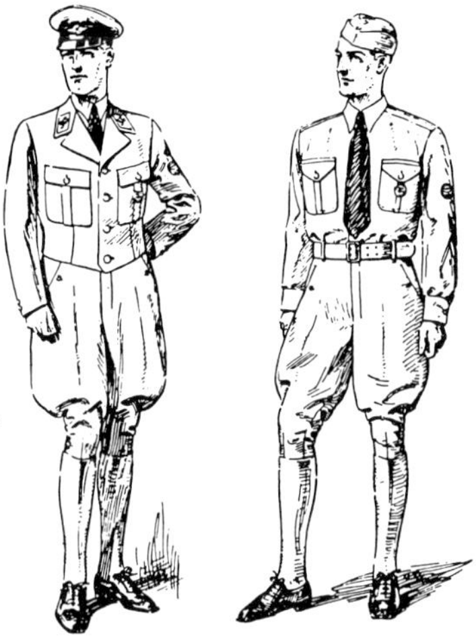

DLV Working-Dress for glider pilots

General der Flieger Fr. Christiansen at the 1935 Deutsch-Landflug. Note glider pilot at his left.

A landesgruppe number insignia was sewn onto the shoulder strap of the "Segelflug" working-dress tunic.

LUFTFAHRT/DLV RANK INSIGNIA (SEPT. 1933-FEB. 1935)

COMMISIONED RANKS

1

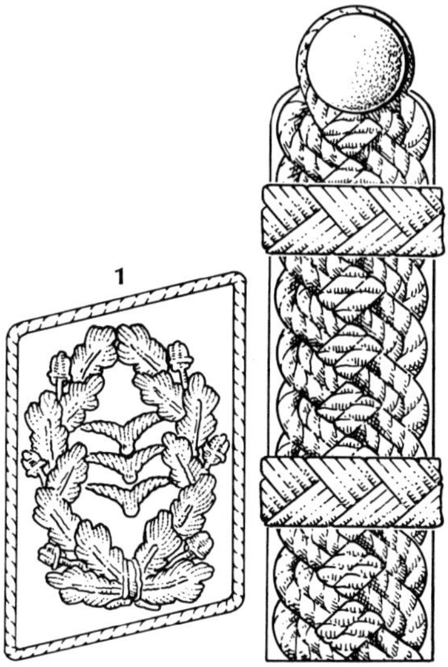

2

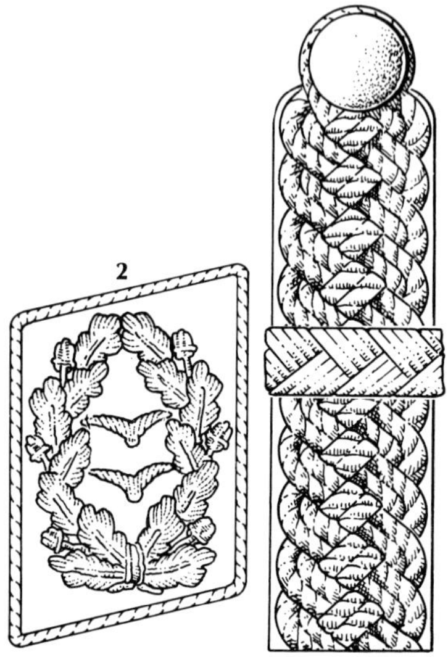

GOLD

3

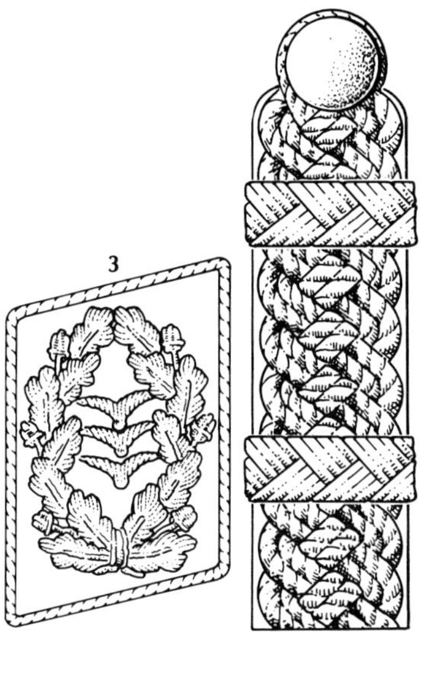

4

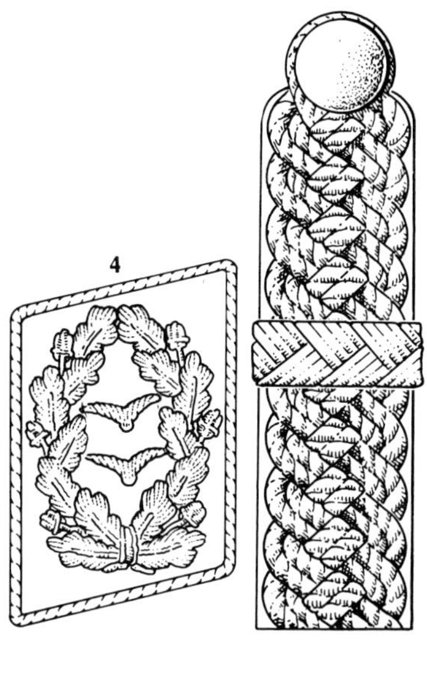

Bruno Loerzer, leader of the Deutscher Luftsport-Verband

5

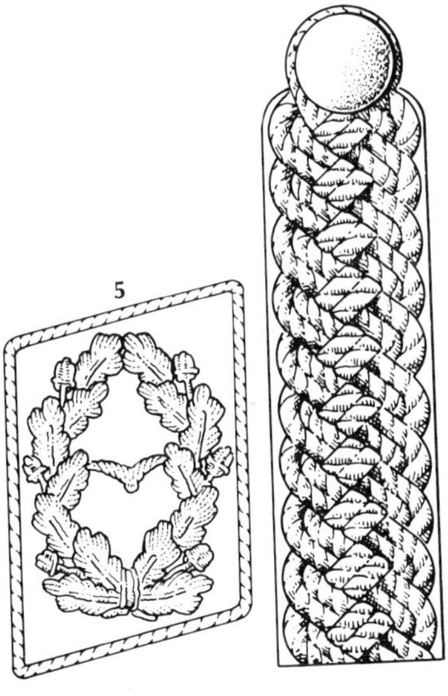

6

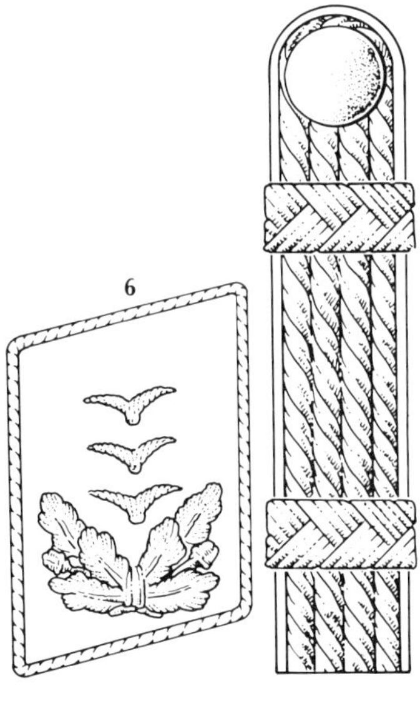

7

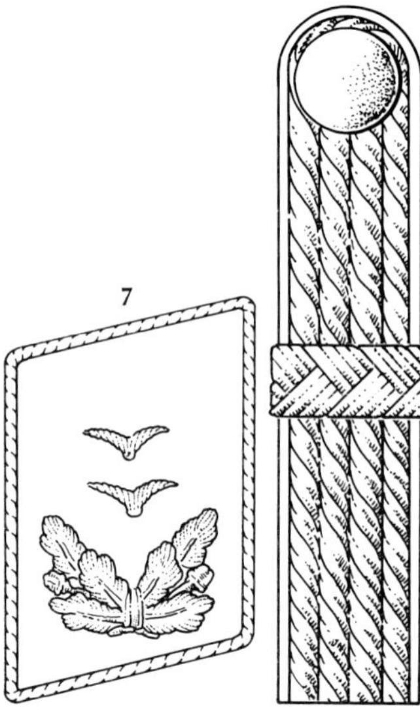

8

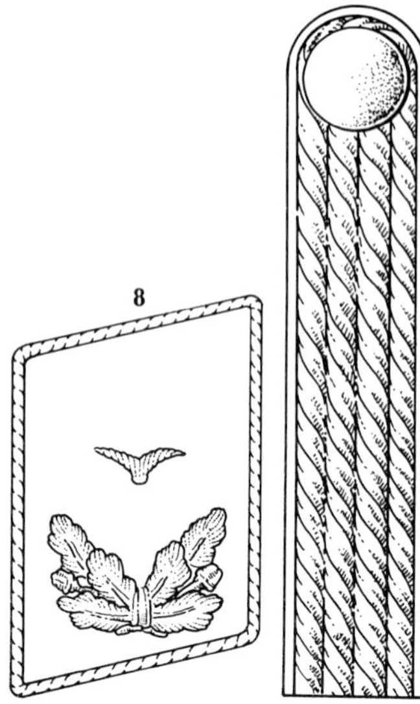

NON-COMMISSIONED RANKS

9

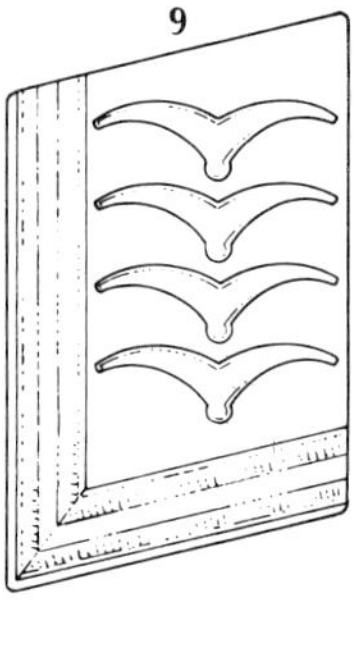

10

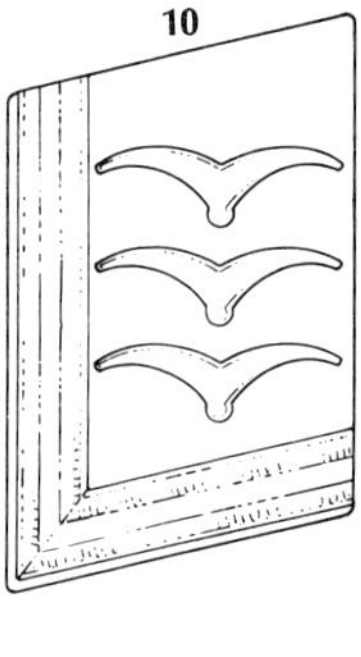

11

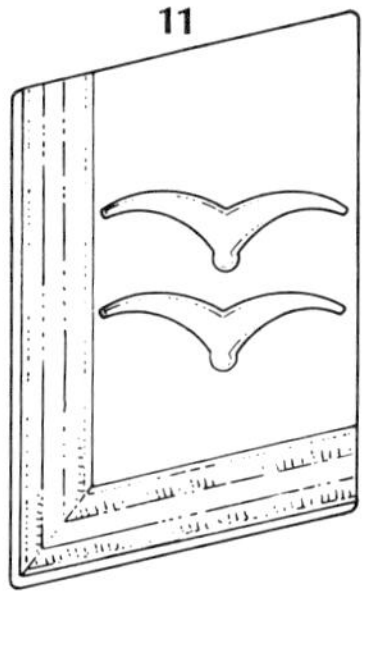

12

1. Reichsminister für Luftfahrt } Gold/Silver/Gold Twist Cord
2. Staatssekretär
3. Fliegerkommodore
4. Fliegervizekommodore
5. Fliegerkommodant
6. Fligerkapitän
7. Schwarmführer
8. Kettenführer
9. Oberflugmeister, Bordoberfunkmeister, Obermeister
10. Flugmeister, Bordfunkmeister, Meister
11. Unterflugmeister, Bordunterfunkmeister, Untermeister
12. 1. Flugzeugführer, 1. Bordfunker, Oberwart
13. 2. Flugzeugführer, 2. Bordfunker, Wart
14. Hilfsflugzeugführer, Hilfsbordfunker, Unterwart
15. Oberflieger
16. Flieger

NOTE:

In "Die Dienstunterricht in Heere" dated 1935, #2 is decribed as Fligerchef (=Generlleutnant) and the collar patch with one wing and no gold slide on the shoulder strap, is described as Fliegervizechef (= Generalmajor). All other references to DLV rank are as illustrated.

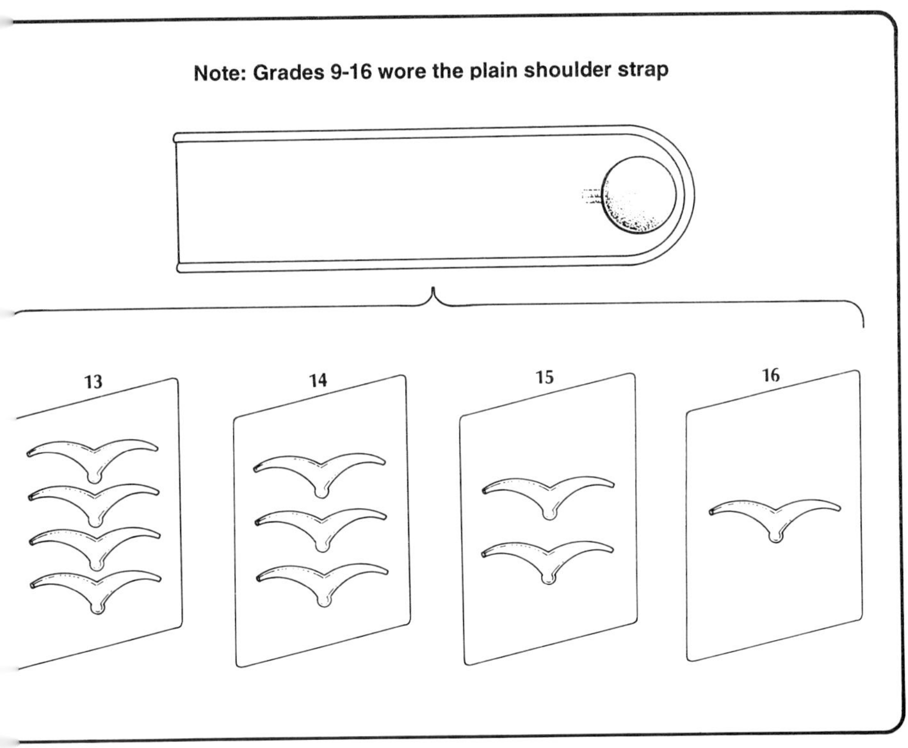

KNIVES AND DAGGERS OF THE DLV

In 1934, the Fligermesser or Flyer's Knife was apporved for wear by all members of the DLV. Its overall length was 34cm, fittings of silver plate over nickel, handle and scabbard covered in blue leather, and a black enamel swastika on a silver circle on the crossguard.

Later that year, a Fliegerdolch or Flyer's Dagger was introduced for wear by DLV officers only. Its design was almost identical to that of the Model 1934 Luftwaffe Dagger with

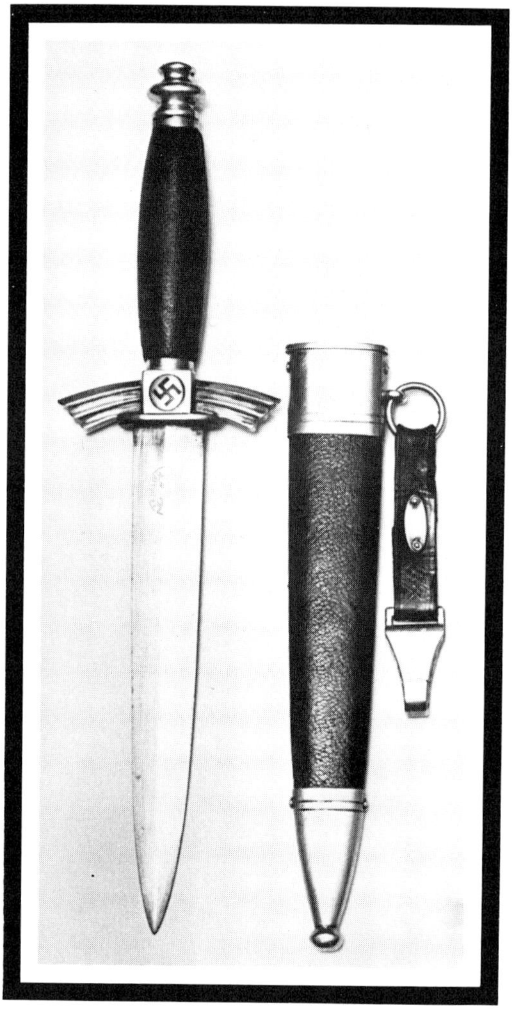

DLV Fliegermesser

Note that the enlisted man at right is wearing the Fliegermesser while the officer is wearing the Fligerdolch.

the exceptions that it was 55cm overall, had no wire wrapping on the handle, and the scabbard had no metal underbody. By mid-1935 this long, cumbersome model was exchanged for the then standard pattern 1934 Fliegerdolch (48cm overall) which had been adopted by the new Luftwaffe.

NOTE:

It should be noted that in 1936 the Fliegermesser was modified by adding 0.5cm to its overall length, fittings were now in dull aluminum and a blue enamel was sprayed over the die-embossed aluminum scabbard, simulating a leather grain. The DLV models (1934 and 1936) are also distinguished by the DLV insignia on the scabbard throat. After April 17, 1937, the newly organized NSFK replaced the DLV and adopted the Fliegermesser as its official sidearm for all ranks. After this date, the NSFK insignia was positioned on the scabbard throat.

Note exceptional length of DLV Fliegerdolch.

MAJOR WICK

Rank Insignia and Waffenfarben

Generalfeldmarschall

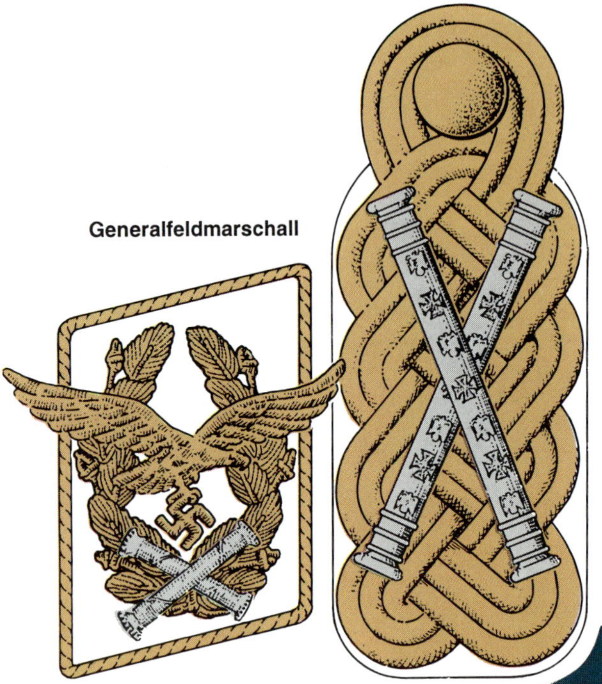

NOTE:
Göring became Generalfeldmarschall on February 4, 1938, with the offical order for the new rank insignia dated February 25, 1938.

Generalfeldmarschall
Sperrle

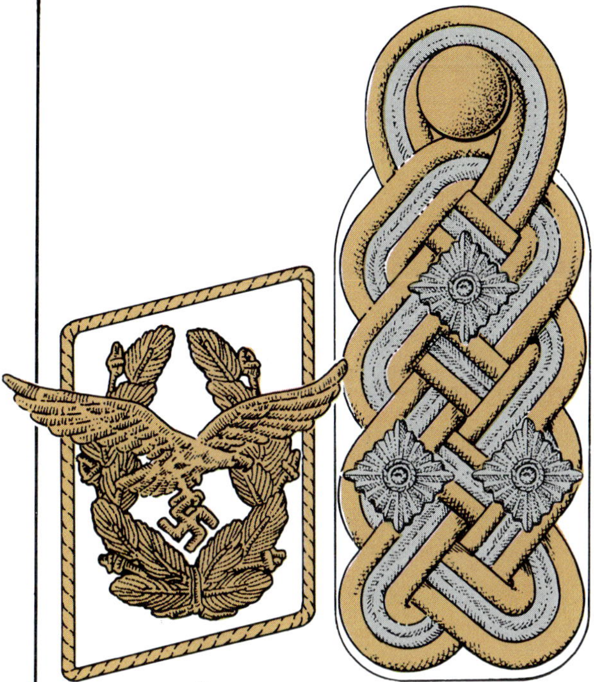

Generaloberst

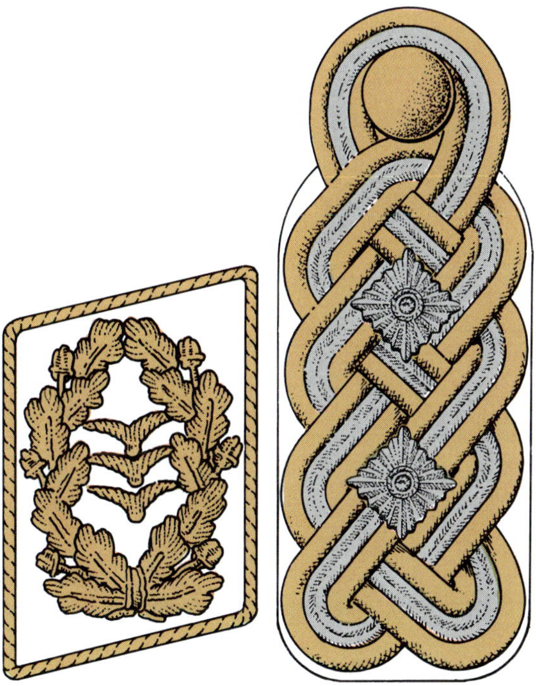

General der Flieger, General der Flakartillerie, General der Luftnachrichtentruppe and General der Fallschirmtruppe

Generalleutnant

Generalmajor

Oberst

Oberstleutnant

NOTE:

All ranks are shown with yellow Waffenfarbe (Fliegertruppe and Fallschirmtruppe). Multiple designations for certain ranks indicate designations for other branches of the Luftwaffe (i.e. Sanitätsoberfeldwebel - Medical Corps).

Major

Hauptmann

Oberleutnant

Leutnant

NOTES:

1. The Tresse around the collar of the Stabsfeldwebel rank below was worn by all NCOs (except Oberfähnrich and Unterarzt who wore officers uniforms with NCO peaked caps and aluminum cap cords). Copper-brown Tresse was worn on Tropical shoulder straps.

2. Hauptfeldwebel/Hauptwachtmeister ranks were by appointment only and were distinguished by two rings of NCO Tresse on both sleeves. When worn with the Afrikakorps cuff title, the cuff title was positioned directly above the Tresse (copper-brown).

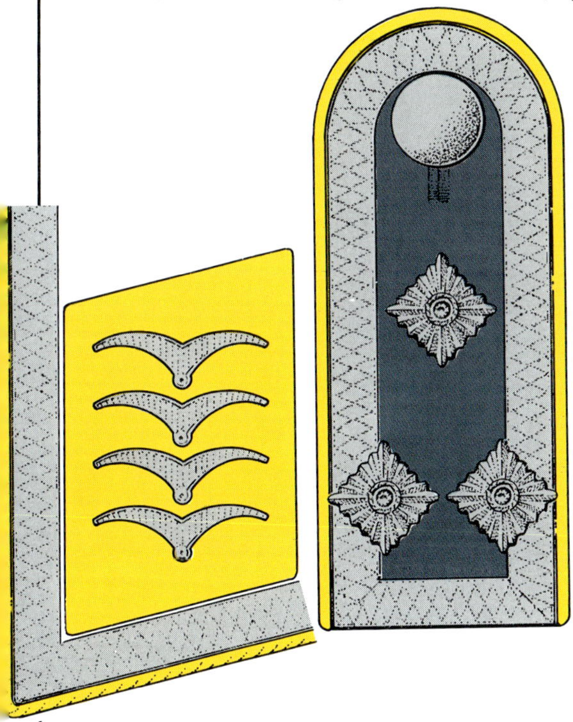

Stabsfeldwebel, Stabswachtmeister, Sanitätsstabsfeldwebel, Stabsfeuerwerker (also appointment as Hauptfeldwebel or Hauptwachtmeister)

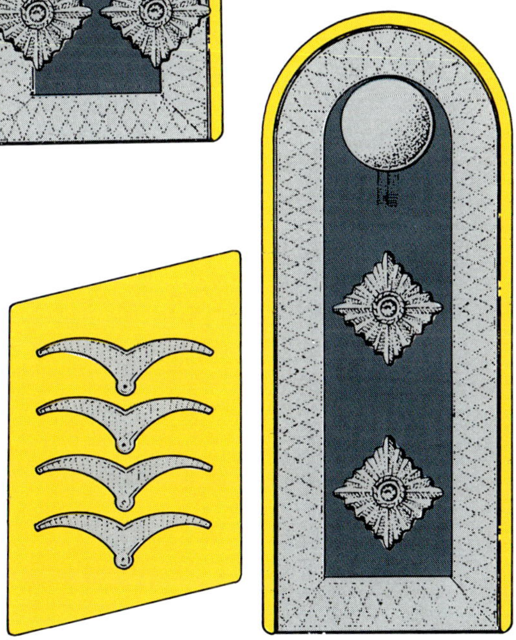

Oberfeldwebel, Oberwachtmeister, Oberfähnrich, Sanitätsoberfeldwebel, Oberfeuerwerker (also appointment as Hauptfeldwebel or Hauptwachtmeister)

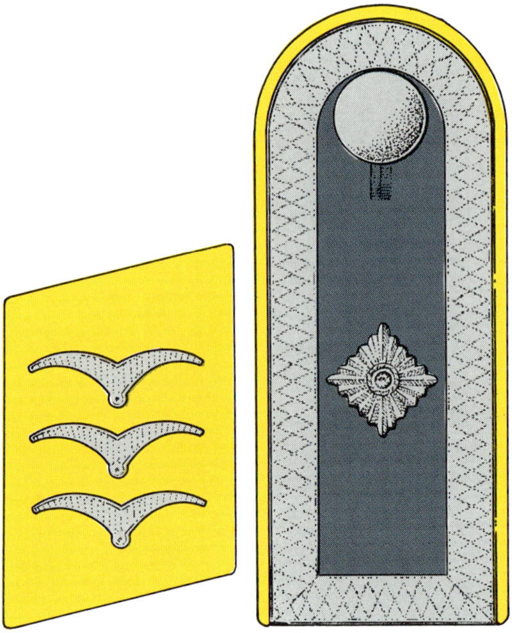

Feldwebel, Wachtmeister,
Sanitätsfeldwebel, Feuerwerker

Feldwebel Schuster

NOTE:
Collar patches with NCO Tresse were worn on great-coats by all NCO ranks.

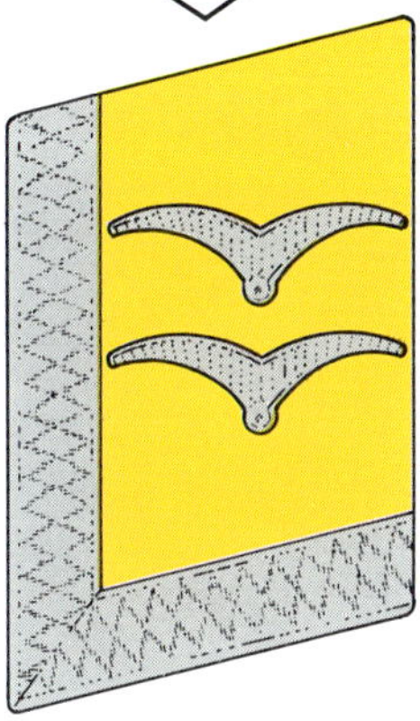
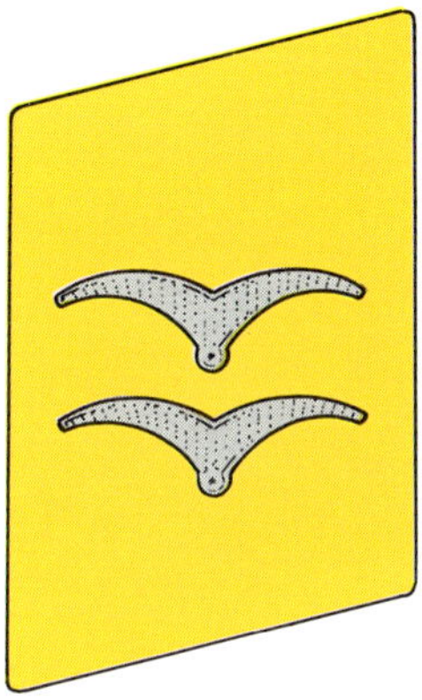
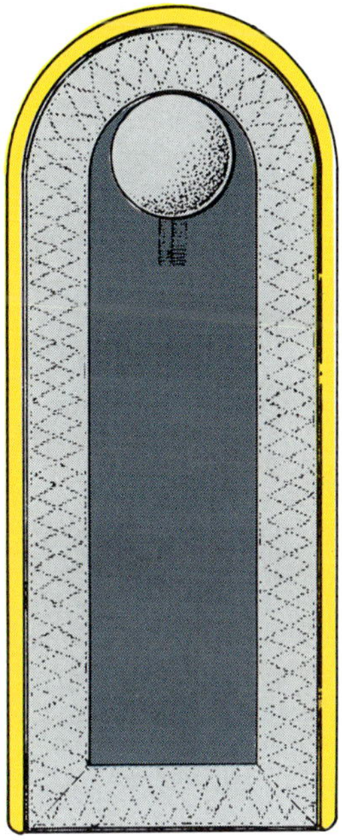

Unterfeldwebel, Unterwachtmeister, Fähnrich, Sanitätsunterfeldwebel

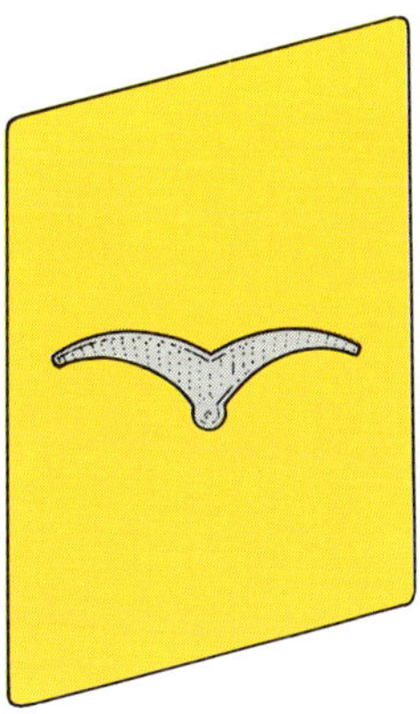
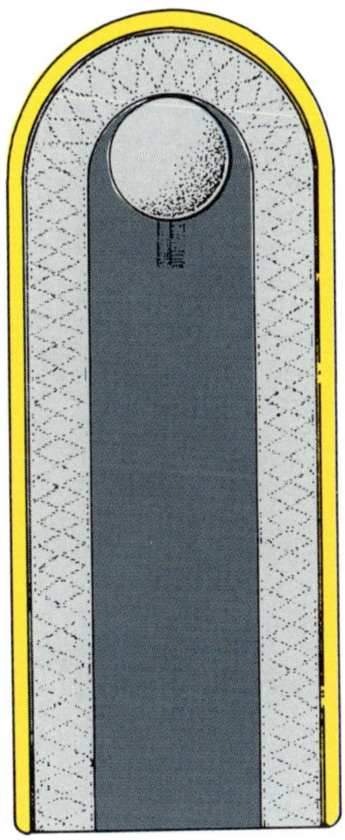

Unteroffizier, Sanitätsunteroffizier, Oberjäger (Fallsch.-Truppe, Jäger- and Wachbataillone Regt. Gen. Göring)

Hauptgefreiter, Sanitätshauptgefreiter

Hauptgefreiter

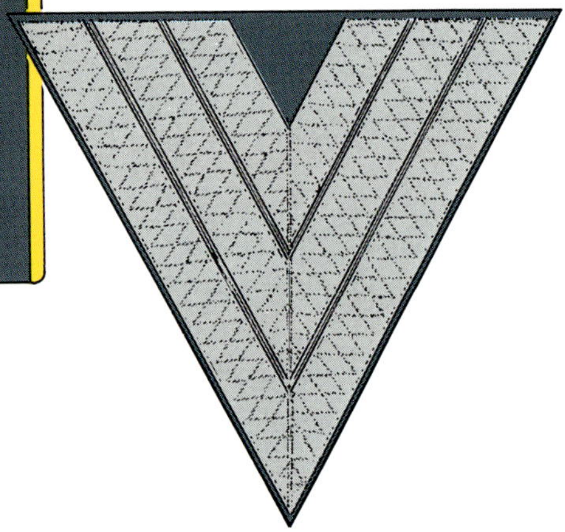

Stabsgefreiter (initiated February 4, 1944)

NOTE:

The rank chevrons for the upper sleeve were initiated for wear on March 4, 1938. After May 12, 1944, the rank of Hauptgefreiter was discontinued and replaced by the new rank designation of Stabsgefreiter. At this time, the Hauptgefreiter rank chevron was replaced by that of the Stabsgefreiter (with woven pip).

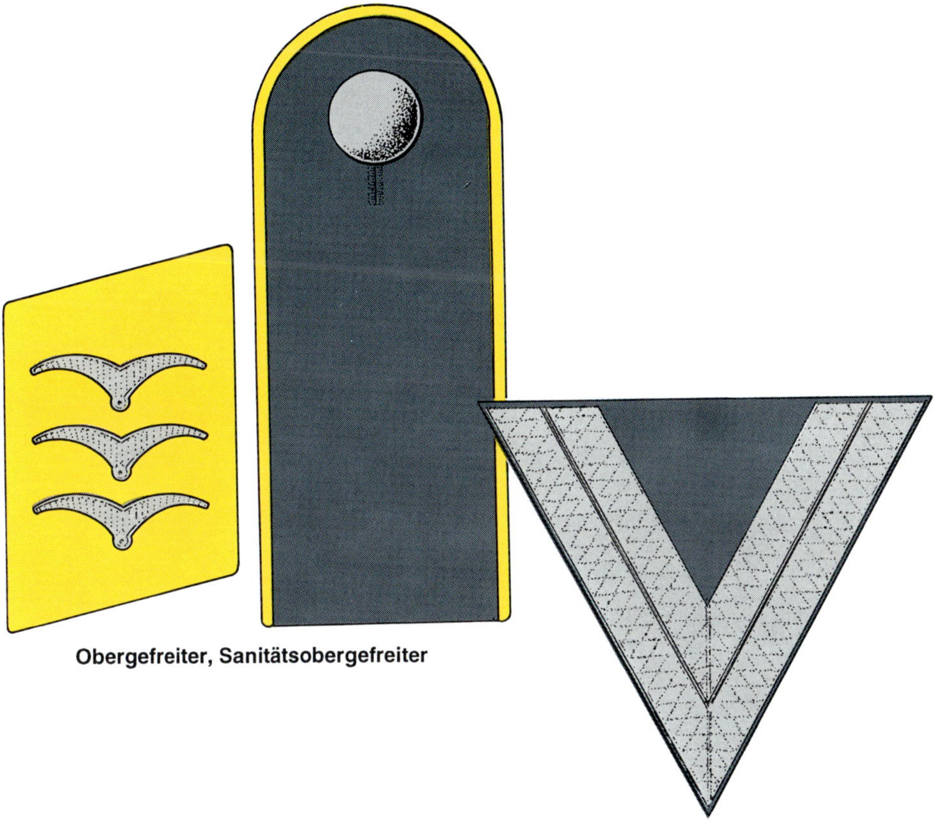

Obergefreiter, Sanitätsobergefreiter

Gefreiter, Sanitäts-
gefreiter

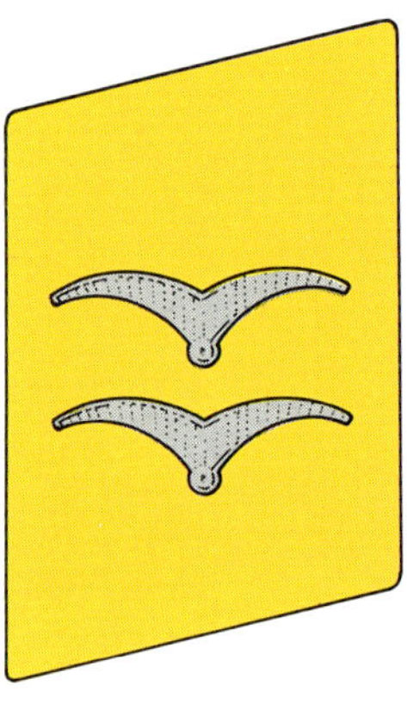

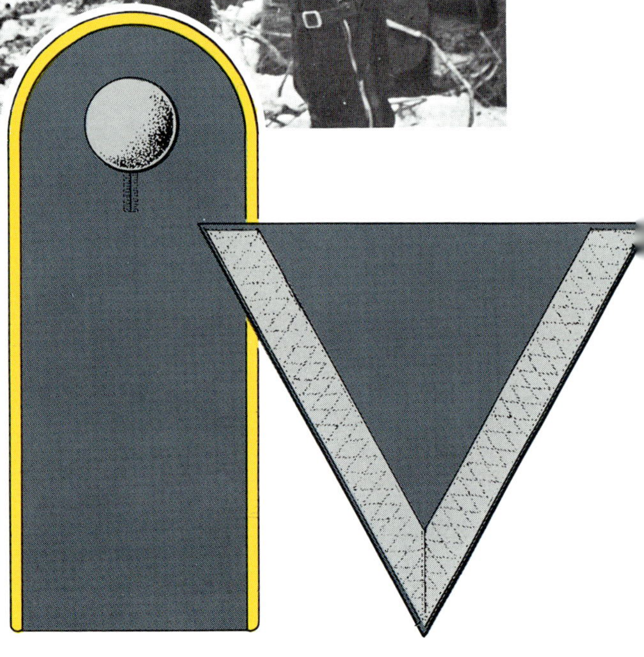

Flieger, Kanonier, Funker, Sanitätssoldat, Jäger (Fallsch. Truppe, Jäger- and Wachbataillone Regt. Gen. Göring)

Corps of Administrative Officials AFTER 1940

Administrative officials (Wehrmachtbeamte) who were in NCO ranks wore the uniform of the Luftwaffe's Feldwebel/Oberfeldwebel with their Waffenfarbe in dark green and triangular pips instead of wings on their collar patches. An inner underlay on their shoulder straps indicated their corps or branch. These officials who were not in an officer's rank did, however, rank higher than the Oberfeldwebel.

Officials in an officers rank wore the officers uniform with their base Waffenfarbe in dark green and embroidered, aluminum triangular pips instead of wings (gold for officials in generals' rank). The inner underlay of their shoulder strap indicated their corps or branch.

CORPS OR BRANCH	SHOULDER STRAP UNDERLAY/ INNER UNDERLAY	COLLAR PATCHES
Officials with rank of general officers (Wehrmachtbeamte im Generalsrang)	Dark Green/White	Dark Green
Court Martial Officials (Obstlt.-Genlt.) (Beamte des Reichskriegsgerichts)	Dark Green/Bordeaux Red	Bordeaux Red
Court Martial Officials (Lt.-Obstlt.) (Beamte des Reichskriegsgerichts)	Dark Green/Bordeaux Red	Dark Green
Judicial Officials (Hptm.-Genmaj.) (Justizdienst)	Dark Green/Wine Red	Wine Red
Judicial Officials (1 - Major) (Justizdienst)	Dark Green/Wine Red	Dark Green
Misc. Technicians (see group 1 below)	Dark Green/Pink	Dark Green
Signals (Lt.-Obstlt.) (Nachrichtendienst)	Dark Green/Golden-Brown	Dark Green
Misc. Specialists (see group 2 below)	Dark Green/Orange	Dark Green
Meteorological (1 - Oberst) (Reichswetterdienst)	Dark Green/Blue-Grey	Dark Green
Air Traffic Control (3a - Lt.) (Reichsluftaufsichtsdienst)	Dark Green/Light Green	Dark Green
Flight Safety (1 - Oberst) (Reichsflugsicherungsdienst)	Dark Green/Dark Brown	Dark Green
Misc. Technicians and Administrators (see group 3 below)	Dark Green/Red	Dark Green
The Engineer, Navigation and Flying Leader Corps have not been included in this chart due to their uniqueness plus war-time changes. They are therefore covered separately in this section.		

GROUP I

Service	Ranks
Photo Service (Bilddienst)	1 – Obstlt.
Motor Transport Service (Kraftfahrdienst)	
Ordnance Service (Waffendienst)	
Ammunition Service (Munitionsdienst)	
Aircraft Technician Service (Flugzeugtechnischer Dienst)	
Cartographic Service (Karthographischer Dienst)	
Lithographic Service (Lithographischer Dienst)	
Craftsmen's Service (Handwerklicher Dienst)	

GROUP II

Specialist School Instructors Service (Fachshuldienst)
Librarian Service (Bibliotheksdienst)
Medical Service (Medizinaldienst)
Pharmacists (Apotheker)
Chemists (Chemiker)
Veterinary Service (Veterinärdienst)
Officials of the Luftwaffe War Science Section (Beamte der Kriegswissenschaftlichen Abteilung der Luftwaffe)

} Lt. – Oberst

GROUP III

General Adminstration Service (Allg. Verwaltungsdienst)
Building Administration Service (Bauverwaltungsdienst)
Fire Protection Service (Feuerschutzdienst)
Agricultural Service (Landwirtshaftsdienst)
Surveying Service (Vermessungsdienst)
Machine Technician Service (Maschinentechnischer Dienst)
Electro-Technical Service (Elektrotechnischer Dienst)
Craftsmen's Services in Air District Clothing Depots (Handwerklicher Dienst bei Bekleidungs-Ämtern)

} 1 – Oberst

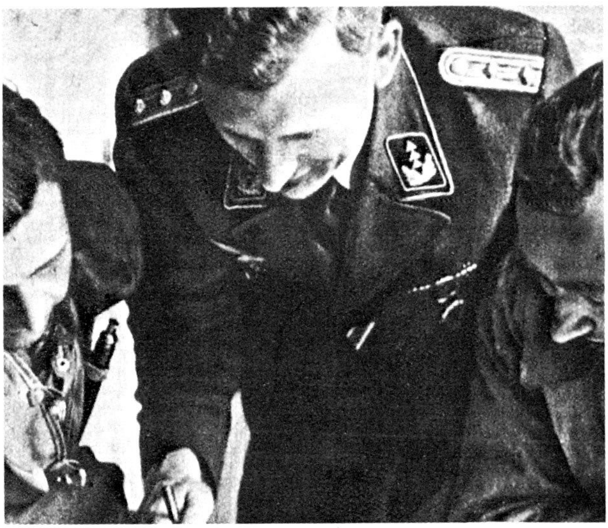

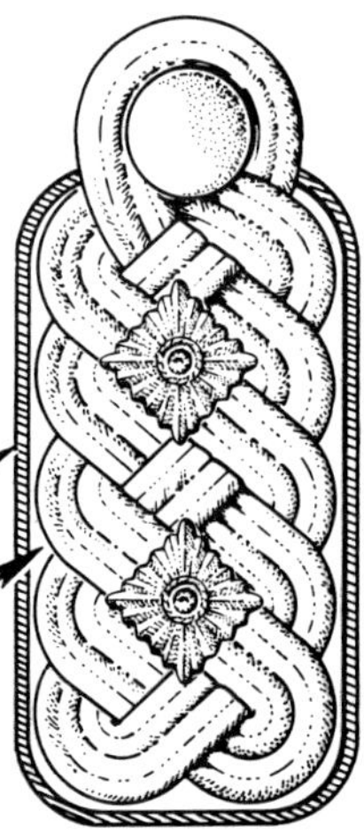

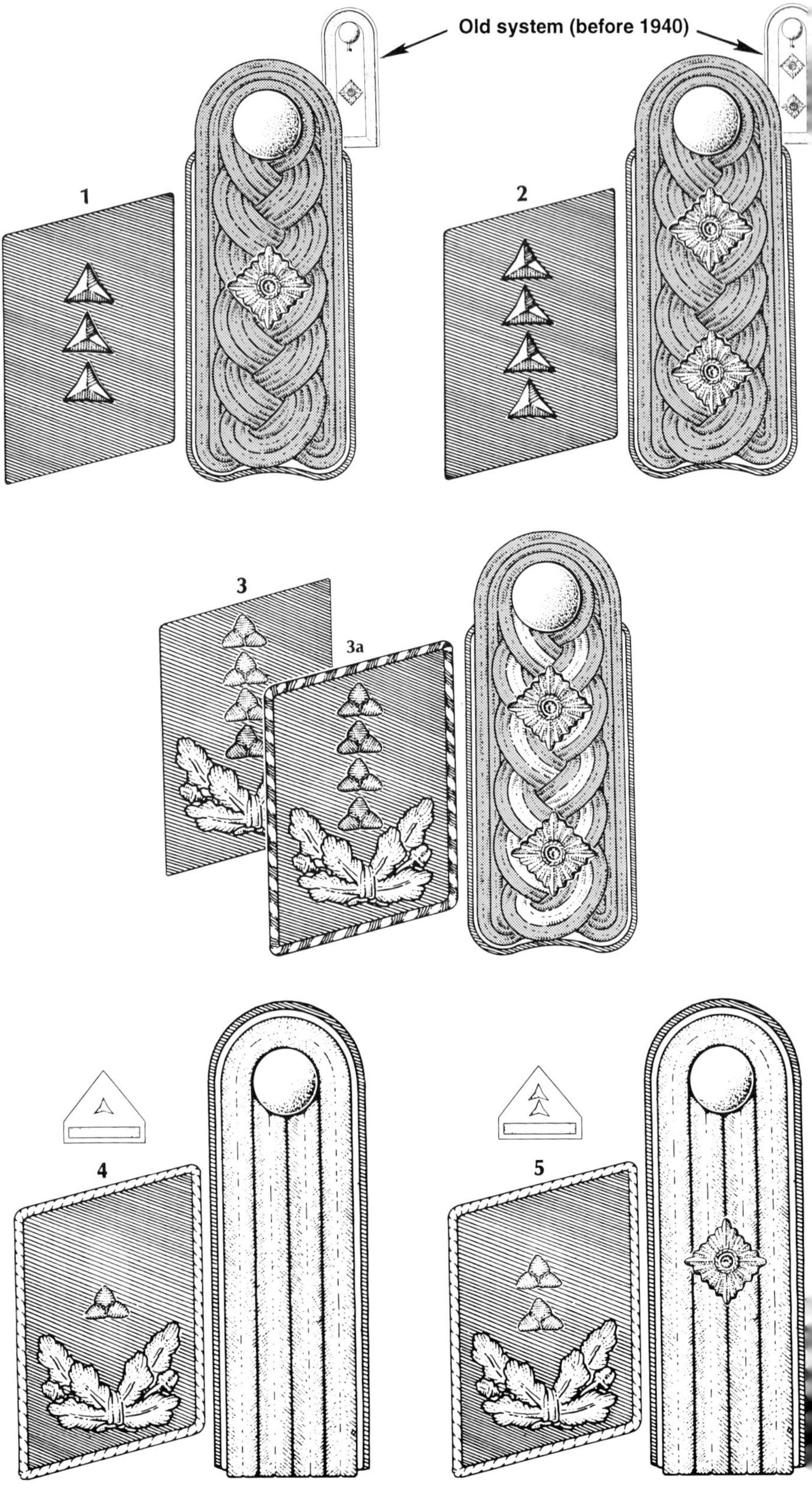
Old system (before 1940)
1
2
3
3a
4
5

NOTE:

Officials rank insignia (#1-11) had Waffenfarben collar patches and shoulder patches and shoulder strap underlays which reflected their corps (see Wehrmachtbeamten color schemes on previous page).

1 & 2: The entwined shoulder cords were in blue-grey with the pips in aluminum. The triangular pips on the collar patches were also in aluminum (einfacher Dienst*).

3 & 3a: The entwined shoulder cords were in blue-grey, with the exception of the middle cord which was in aluminum. The collar patches were piped with an alternating aluminum/dark green/aluminum cord (old system aluminum only) for (3a) "Beamte des mittleren Dienstes"** and had no piping for (3) "Beamte des einfachen Dienstes."* Shoulder strap pips were in aluminum and collar patch pips in matt-grey embroidery.

4, 5 & 6: The flat shoulder cords were in aluminum (as for Leutnante u. Hauptleute) and their pips were in a gold color. Embroidery on the collar patches was in aluminum wire. Piping around the collar patches was in gold for "Beamte des höheren Dienstes"****, in aluminum for "Beamte des gehobenen Dienstes"*** and in alternating aluminum/dark green/aluminum for "Beamte für mittleren Dienstes."

7, 8 & 9: The entwined shoulder strap cords were in aluminum and their pips in a gold color. Embroidery on the collar patches was in aluminum wire. Piping around the collar patches was in gold for "Wehrmachtbeamte des höheren Dienstes" and in aluminum for "Wehrmachtbeamte des gehobenen Dienstes."

10 & 11: The entwined shoulder strap cords were for general officers, with two gold colored cords and one aluminum colored cord (in the middle). Pips on the shoulder straps were in aluminum. Embroidery and piping on the collar patches were in gold. The shoulder strap inner underlay was white until 25.4.44. From that date, it was in the appropriate branch color (höherer Dienst).

* "Beamte des einfachen Dienstes" – An individual with a grade school education.

** "Beamte des mittleren Dienstes" – Generally an individual with a high school education (he may have taken some additional schooling, including military).

*** "Beamte des gehobenen Dienstes" – An individual who had received a formal education but not necessarily university level (generally a specialized education . . . engineering, law, etc.).

**** "Beamte des höheren Dienstes" – A university graduate.

6

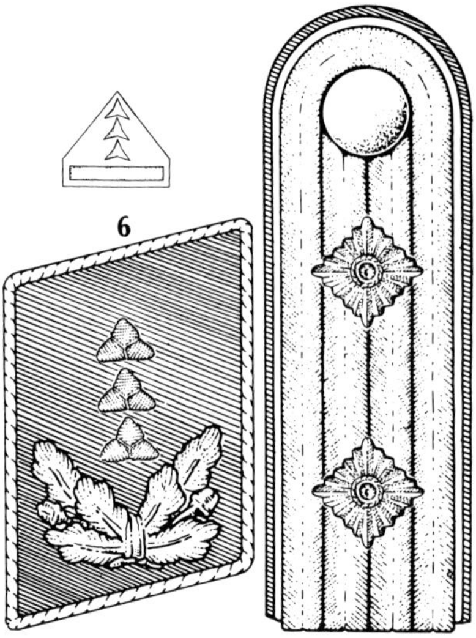

7

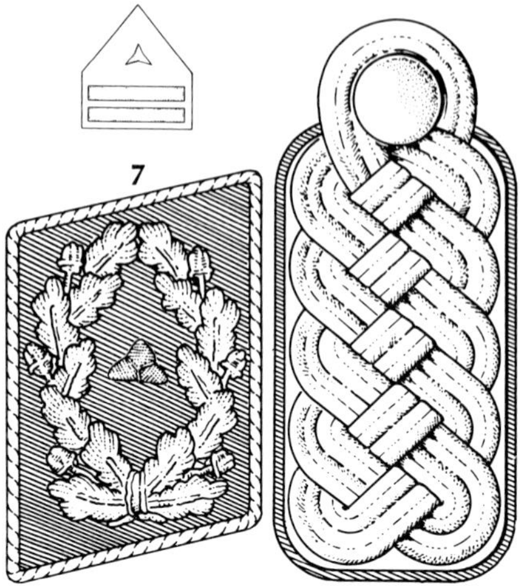

8

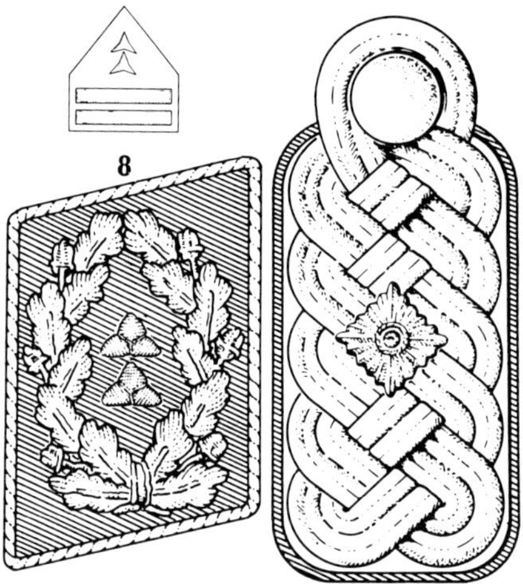

9

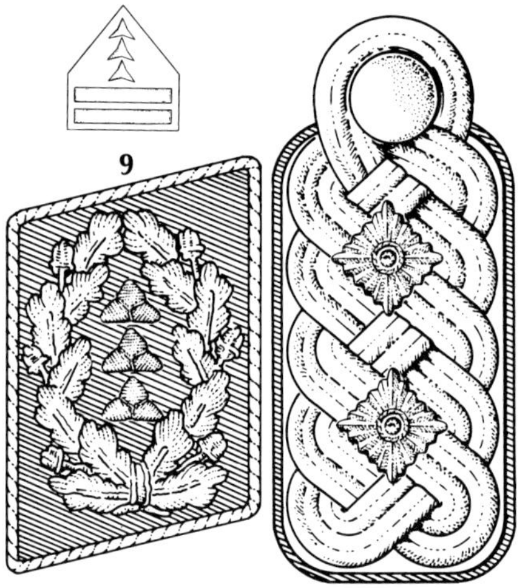

10

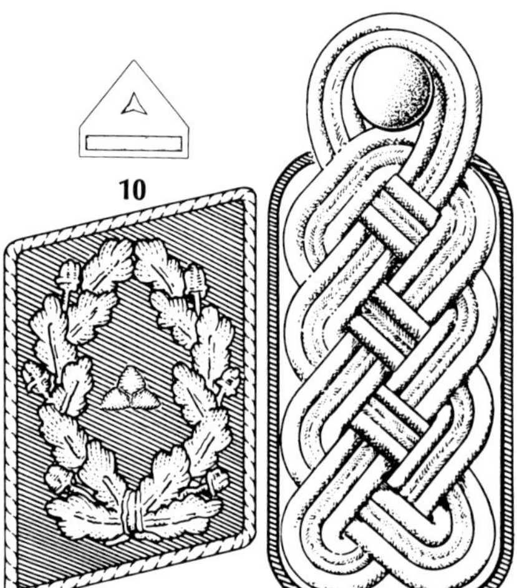

11

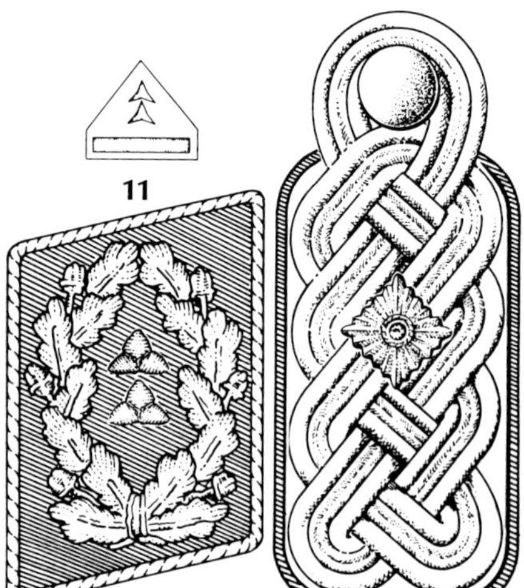

By using the Wehrmachtbeamten Waffenfarben chart plus the numbered rank illustrations, any rank designation from any corps or branch can be calculated. The numbers preceding them following rank designations are coded to the insignia illustrations.

COURT MARTIAL OFFICIALS (OBSTLT.-GENLT.)

8. Oberkriegsgerichtsrat beim R.K.G. (Reichskriegsgericht)****
9. Oberskriegsgerichtsrat beim R.K.G.****
10. Reichskriegsgerichtsrat/Reichskriegsanwalt**** } pant stripes & great-coat lapels in bordeaux red
11. Senatspräsident/Oberreichskriegsanwalt**** } pant stripes & great-coat lapels in bordeaux red

COURT MARTIAL OFFICIALS (LT.-MAJOR)

4. Reichskriegsgerichtssekretär**
5. Reichskriegsgerichtsobersekretär/ Reichskriegsgerichtsinspektor***
6. Reichskriegsgerichtsoberinspektor***
7. Amtmann u. Amstrat beim R.K.G.***
8. Bürodirektor beim R.K.G.****

JUDICIAL OFFICIALS (HAUPTM.-GENMAJ.)

6. Kriegsrichter****
7. Kriegsgerichtsrat der Luftwaffe****
8. Oberkriegsgerichtsrat der Luftwaffe****
9. Oberstkriegsgerichtsrat der Luftwaffe/Ministerialrat****
10. Ministerialdirigent/Lw.-Chefrichter**** } pants stripes & great-coat lapels in wine red

JUDICIAL OFFICIALS (1.-MAJOR)

1. Justizwachtmeister*
2. Justizoberwachtmeister*
4. a.p. Kriegsgerichtsinspektor***
5. Kriegsgerichtsinspektor***
6. Kriegsgerichtsoberinspektor***
7. Amtmann u. Amstrat beim Luftwaffengericht***

MISC. TECHNICIANS (SEE GROUP I) (1.-OBERSTLEUTN.)

1. Betriebsassistent*
2. Betriebsassistent (after 5 years service as same)*
3a. techn. Assistent/Prüfmeister/Werkmeister**
4. Oberwerkmstr./Oberprüfmstr./Betriebsleiter/techn. Sekretär/** a.p.techn. Inspektor***
5. techn. Inspektor***
6. techn. Oberinspektor***
7. techn. Amtmann/techn. Oberamtmann/Amtsrat***
8. Amtsrat (after special appointment)***

SIGNALS OFFICIALS (LT.-OBERSTLEUTN.)
4. a.p. techn. Inspektor***
5. techn. Inspektor***
6. techn. Oberinspektor***
7. techn. Amtmann/techn. Oberamtmann/Amtsrat/Oberamtmann***
8. Amtsrat (after special appointment)

MISC. SPECIALISTS (SEE GROUP II) (LT.-OBERST.)
4. Feldapotheker****
5. Regierungs-Inspektor/a.p. Oberfahnschullehrer***/Oberapotheker****
6. Reg.-Oberinspektor/Fachschulkonrektor/Gewerbeoberlehrer/ Oberfachschullehrer***/ Fachstudienassessor/Stabsapotheker****
7. Amtmann/Amtsrat/Fachschulrektor***/Regierungsrat/ Fachstudienrat/Fachstudiendirektor/Oberstabsapotheker/Regierungsm edizinalrat/Regierungschemierat****
8. Amtsrat (after special appointment)/Oberfachschulrat***/ Oberreg.rat/Fachstudiendir./ Oberfeldapotheker/ Oberregierungsmedizinalrat/Oberregierungschemierat****
9. Ministerialrat****

METEOROLOGICAL OFFICIALS (1.-OBERST)
1. Betriebassistent*
2. Betriebsassistent (after 5 years as same)*
3a. Wetterdienst-Assistent**
4. Wetterdienst-Sekretär/a.p. techn. Inspektor***
5. techn. Inspektor/Wetterdienst-Obersekretär
6. techn. Oberinspektor***/Wetterdienstassessor****
7. techn. Oberamtmann***/Regierungsrat****
8. Amtsrat (after special appointment)***/Oberregierungsrat
9. Ministerialrat/Direktor im Reichswetterdienst/Direktor b.d. Seewarte/Direktor b. Reichsamt für Wetterdienst****

AIR TRAFFIC CONTROL OFFICIALS (3a-LT.)
3a. Untermeister im Reichsluftaufsichtsdienst*
4. Meister im Reichsluftaufsichtsdienst**

FLIGHT SAFETY OFFICIALS (1.-OBERST)
1. Betriebsassistent*
2. Betriebsassistent (after 5 years service as same)*
3a. techn. Assistent**
4. a.p. techn. Inspektor***
5. techn. Inspektor***
6. techn. Oberinspektor***
7. techn. Amtmann/techn. Oberamtmann/techn. Amtsrat***/Reg.-Baurat****
8. Amtsrat (after special appointment)***/Oberreg.-Baurat****
9. Ministerialrat****

MISC. TECHNICIANS AND ADMINISTRATORS (SEE GROUP III) (1.-OBERST)

1. Ministerialamtsgehilfe/Amtsobergehilfe/Drucker/Bauaufseher/Labora nt/Lagermeister/ Betriebsassistent/Materialienverwalter/Magazinauf seher/Amtsgehilfe/(after 5 years service, previous titles as 2) . . . Heizer/Hauswart/Lagerwart/Wächter/(after 10 years service, previous titles as 2) . . . Kastellan/techn. Gehilfe/Pförtner/Kassengehilfe*

2. Machinenmeister in Bes.Gr.A9/Magazinmeister/Botenmeister/those Beamten with 5 years service in the same job, Bes.Gr.A10/Those Beamten with 10 years service in same job, Bes.Gr.A11

3. Maschinenmeister in Bes.Gr.A8/Oberlotsenmeister/Ministerialinspektor*

3a. techn. & nichttechnische Assistent/Amtsmeister/Unterbrandmeister/a.p. Beamte des mittleren Dienstes**

4. techn. * nichttechnischer Sekretär und Kanzlei-Vorsteher in Bes.Gr.A7/Brandmeister**/ a.p. Beamte des gehobenen Dienstes/Zahlmeister (after 3.5.43)***

5. Hauptbrandmeister/Oberbrandmeister**/Ministerialregistrator/Regie rungsinspektor/ Regierungsbauinspektor/Verwaltungsinpektor/Oberzah lmeister (after 3.5.43)***

6. Oberinspektoren der Beamtenlaufbahnen (Regierungsoberinspektor/Regierungsoberbauinspektor/Verwaltungsob erinspektor)/Stabszahlmeister (after 3.5.43)***/ Regierungsassessor/Regierungsbauassessor****

7. Amtmann/Oberamtmann/Amtsrat/Oberstabszahlmeister and Oberfeldzahlmeister (after 3.5.43)***/Regierungsrat/Regierungsbaurat****

8. Amtsrat (after special appointment)***/Ministerialbürodirektor/Oberregierungsrat/Oberr egierungsbaurat/Oberfeldintendant (after 3.5.43)****

9. Ministerialrat/Oberstintendant/Baudirektor der Luftwaffe****

NOTES:

1. An order dated November 29, 1941, set forth the following insignia change:

before 11.29.41
Amtmann/
Oberamtmann/
Amtsrat/***

after 11.29.41(1)*
Amtmann***

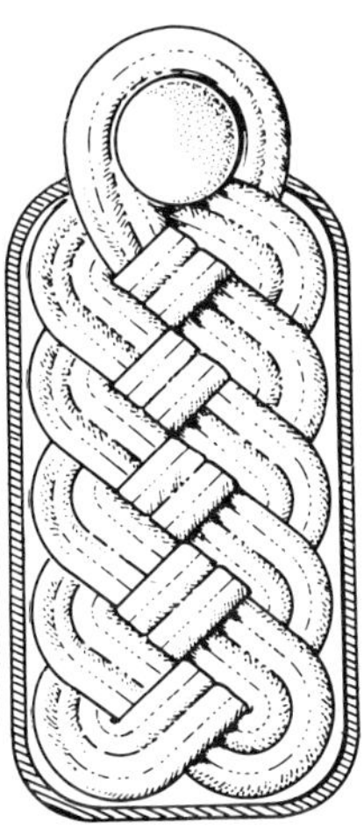

after 11.29.41
Oberamtmann/
Amtsrat***

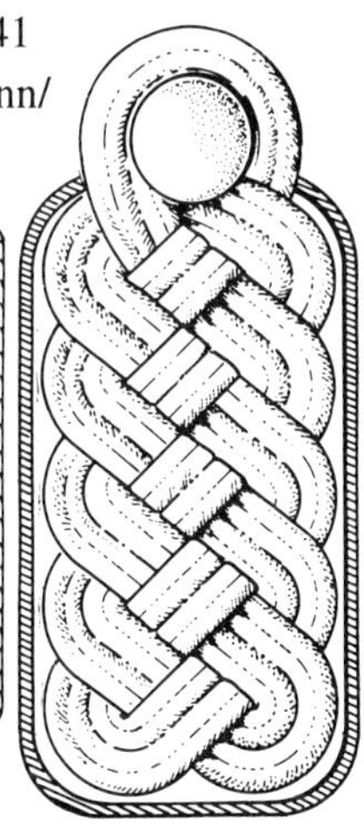

NOTES: (cont.)

The Amtsrat rank (after special appointment) remained unchanged by this order . . . it still had two embroidered triangular pips on the collar patches (as the Oberamtmann/Amtsrat) and a gold colored pip on the shoulder straps (see #8).
2. By an order dated March 26, 1942, Wehermachtbeamte on probation for mittleren or gehobenen Dienst, were to wear the collar patches and shoulder straps as #3a or #4 respectively (p.42) but with a dark green slide on the base of each strap.
3. Officials of the General Administration Service (see Group 3) and of the Court Martial and Judicial Services became "Offiziere im Truppensonderdienst" (TSD) by an order dated May 9, 1944.[2]

TSD/VERWALTUNGSDIENST

4. Zahlmeister
5. Oberzahlmeister
6. Stabsintendant
7. Oberstabsintendant
8. Oberfeldintendant
9. Oberstintendant
10. Generalintendant
11. Generalstabsintendant
12. Generaloberstabsintendant*

* as for General der Flieger, etc. (i.e. three wings on the collar patch and two aluminum pips on the shoulder straps)

Administrative Specialist' insignia (Gold/General Officers –silver)

a. Wings instead of triangular pips were now worn on the collar patches by these services.
b. Also, the administration specialist's insignia was now worn on the shoulder straps by all ranks.
c. Branch color now light blue for all ranks (4.-12.) with the pants stripes for generals in white.

TSD/WEHRMACHTRICHTER

6. Stabsrichter
7. Oberstabsrichter
8. Oberfeldrichter
9. Oberstrichter
10. Generalrichter
11. Generalstabsrichter

a. Wings instead of triangular pips were now worn on the collar patches and the judicial sword insignia was worn on the shoulder straps by all ranks (gold/general officers – silver).
b. Branch color now wine red for all ranks (6.-11.) with the pants stripes for generals in white.

Judicial Sword insignia (Gold/ General Officers – Silver)

4. By an order dated August 19, 1944, the officials of the Building Administration Service (Bautechnisches Verwaltungswesen) received a Waffenfarbe change from red to black. From this date, black was the inner underlay of the shoulder strap with the basic underlay still in green.[3] Collar patches were now in black (no longer dark green) with triangular pips. Officials in general's ranks also carried the black collar patches, black double trouser stripes and black lapels on the great-coat.

Oberregierungsbaurat rank insignia (Building Administration Service)

Flying Leader Corps

Officials of the Flying Leader Corps (Flugführerkorps) were highly specialized Luftwaffe officials who controlled operations within a particular area (defense, supply, reconnaissance, etc.) and were directly responsible to their local Luftflotte commander. Initially, they wore the officials' uniform with dark green collar patches (aluminum piping and triangular pips) and shoulder straps with a dark green underlay and a secondary underlay of golden-yellow. By an order dated October 8, 1941, Flying Leader Corps officials now wore pink collar patches with propeller (as for Engineer and Navigation Corps) and shoulder straps with a pink underlay and a secondary underlay of golden-yellow. The ranks were as follows:

(a) Flugführer (Leutnant) (4)
(b) Oberflugführer (Oberleutnant) (5)
(c) Hauptflugführer (Hauptmann) (6)
(d) Stabsflugführer & Stabsflugführer I.Klasse (Major) (7)

[1] Luftwaffen-Verordungsblatt, December 8, 1941. This order was numbered 1978 and dated November 29, 1941.

[2] Luftwaffen-Verordungsblatt, May 15, 1944. This order was numbered 607 and dated May 9, 1944.

[3] Luftwaffen-Verordungsblatt, September 4, 1944. This order was numbered 1189 and dated August 19, 1944.

Secret Field-Police

The officials in the Luftwaffe's Secret Field-Police (geheime Feldpolizei der Luftwaffe) wore the uniform of the Wehrmachtbeamten. The GFP officials had the dark green branch color (collar patches and shoulder strap underlay) plus a second color of wine red (inner underlay on the shoulder straps). Gold-colored GFP metal insignia were positioned on the center of the shoulder straps (all ranks).[4]

Feldpolizeisekretär (shoulder straps as Leutnant (mittlerer Dienst))
Feldpolizeiinspektor (shoulder straps as Oberleutnant (gehobener Dienst))
Feldpolizeikommissar (shoulder straps as Hauptmann (gehobener Dienst))
Feldpolizeidirektor (shoulder straps as Major (höherer Dienst))
Oberfeldpolizeidirektor (shoulder straps as Oberstleutnant (höherer Dienst))

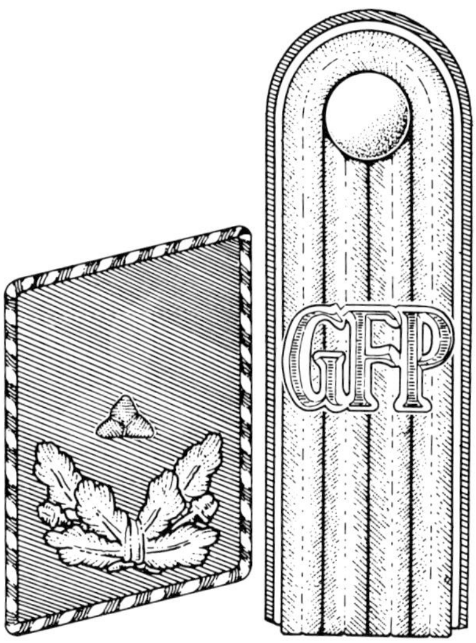

Feldpolizeisekretär

Feldpolizeiinspektor

NOTE:
The principal missions of the GFP were to prevent sabotage and espionage, graft, damage to Luftwaffe property, undermining of moral, and as a fact-finding organ of the court-martial courts.

Feldpolizeikommissar

Feldpolizeidirektor

Oberfeldpolizeidirektor

NOTE:

"Beamte des mittleren Deinstes" – aluminum/dark green/ aluminum collar patch piping

"Beamte des gehobenen Dienstes" – aluminum/collar patch piping

"Beamte des höheren Dienstes" – gold collar patch piping

[4] Luftwaffen-Verordungsblatt, November 15, 1943. This order was numbered 2083 and dated November 5, 1943.

Engineer Corps

Members of the Engineer Corps of the Luftwaffe (Ingenieurkorps der Luftwaffe) were not part of the Corps of Administration and wore a uniform with pink Waffenfarbe and rank variations. Engineers who were on a reserve status, however, wore a light blue border on the pink collar patch plus an inner shoulder strap underlay of light blue. Insignia for ranks up to Fl.-Oberstingenieur were in aluminum embroidery plus gold colored pips on the shoulder straps. The Fl.-Generalingenieur and Fl.-Generalstabsingenieur ranks had gold embroidery on the collar patches, alternate gold/aluminum/gold braid shoulder straps and aluminum piping.

By an order dated November 2, 1940, the title of Ingenieurkorps der Luftwaffe was to be changed to Ingenieuroffizierkorps der Luftwaffe, with the Ingenieurkorps being officially disbanded. This did not occur, however, and for that reason both Corps titles co-existed for the duration of the war. The uniforms and pink Waffenfarbe remained unaltered, with the exception of the collar patches. The collar patches were now in the same pattern as that for the regular Luftwaffe ranks (i.e. embroidered wings instead of propellers). The rank titles for engineering officers were from this date standard Luftwaffe titles, but with an engineer designation (i.e. Leutnant (Ing.), etc.).

In early 1940, the designations of the top three engineering ranks were changed as follows:

Fl.-Oberstingenieur – till now Fl.-Hauptstabsingenieur
Fl.-Generalingenieur – till now Fl.-Chefingenieur
Fl.-Generalstabsingenieur – till now Leitender Fl.-Chefingenieur

Fl.-Genralstabsingenieur

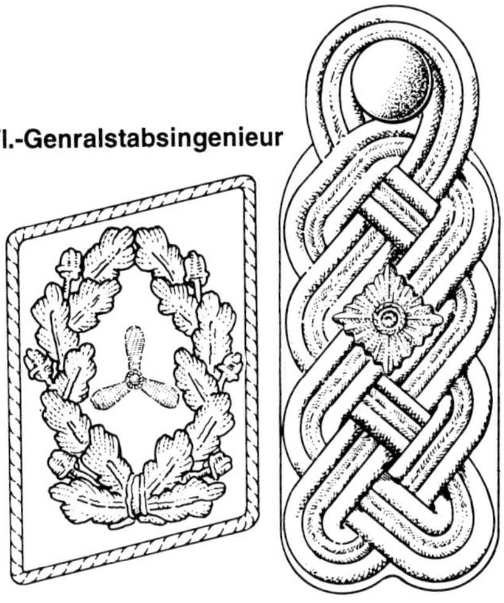

Fl.-Generalingenieur

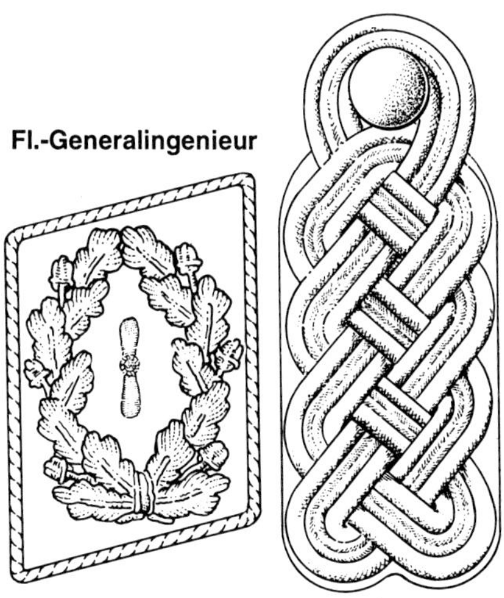

Fl.-Oberstingenieur

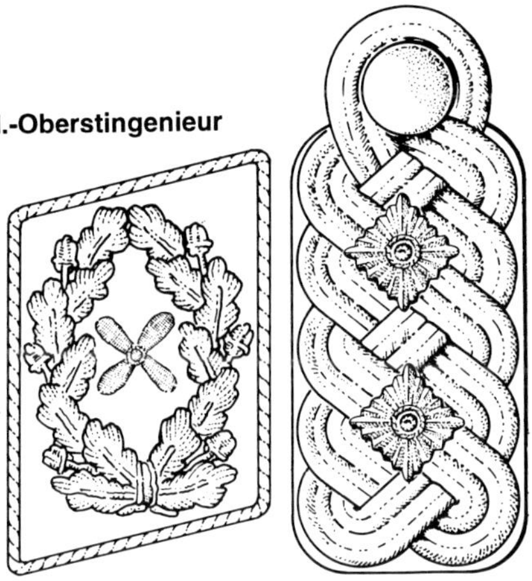

Fl.-Oberstabsingenieur

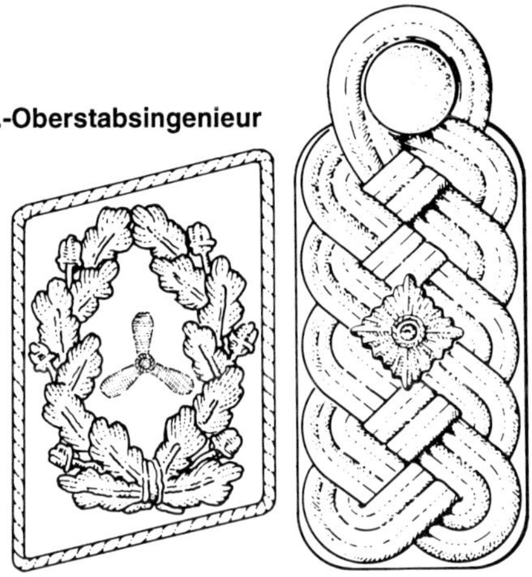

Fl.-Stabsingenieur

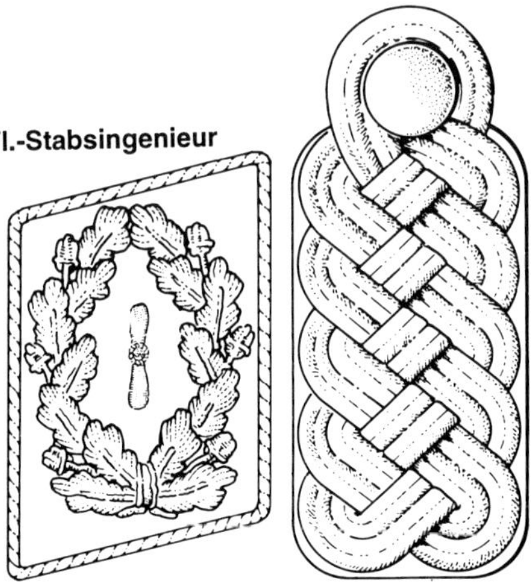

Fl.-Hauptingenieur

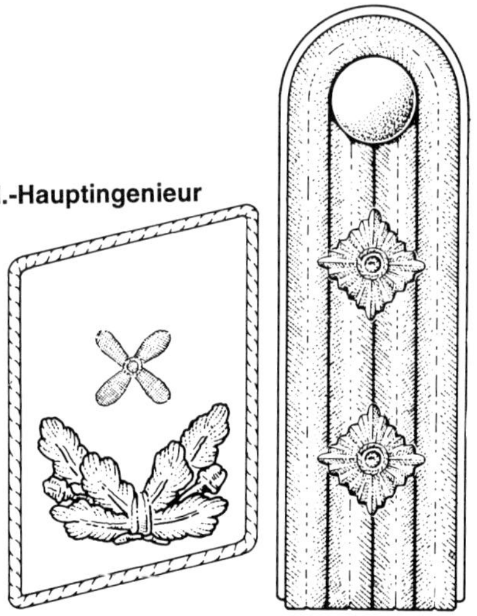

Fl.-Oberingenieur

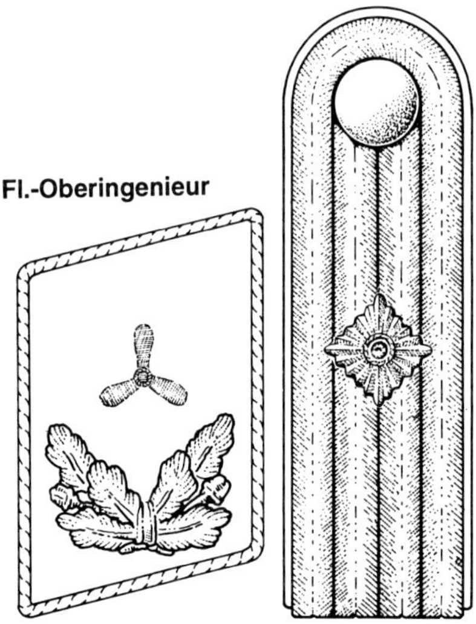

Fl.-Ingenieur

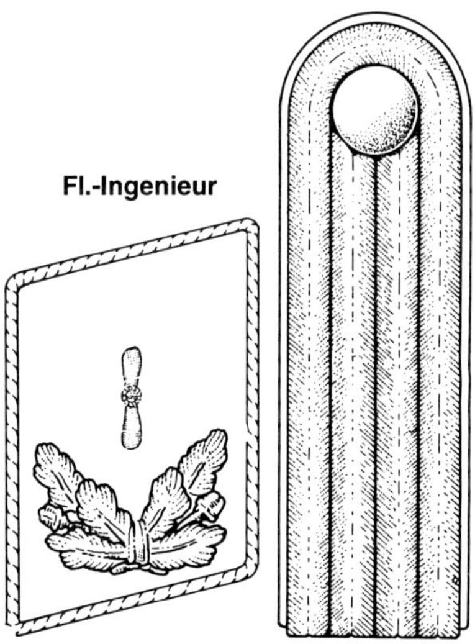

NOTE:

Fliegeringenieuranwärter insignia consisted of a pink slide on the base of the Fl.-Ingenieur shoulder strap (order dated 5.1.43).

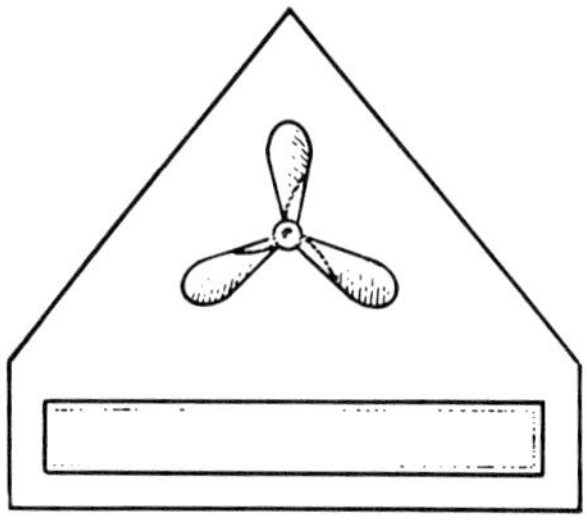

Fl.-Generalstabsingenieur

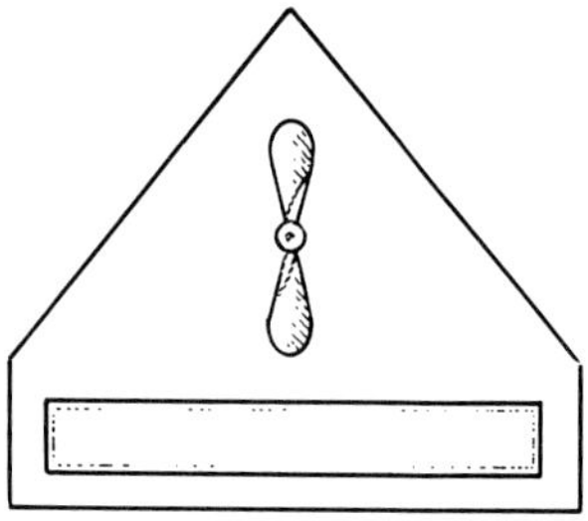

Fl.-Generalingenieur

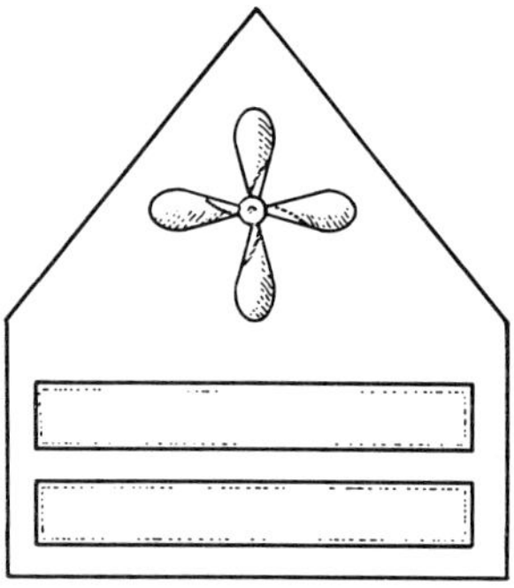

Fl.-Oberstingenieur

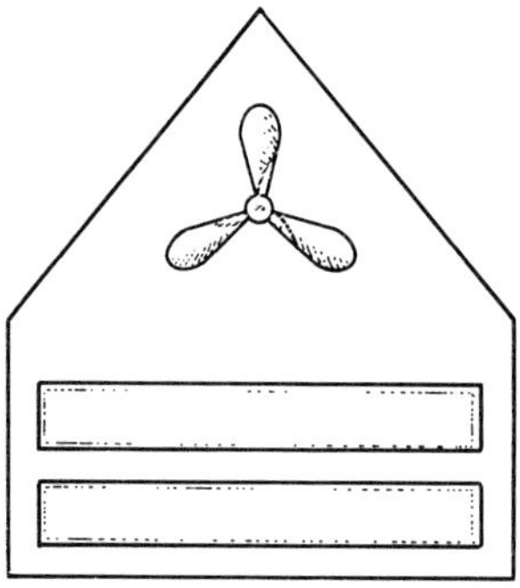

Fl.-Oberstabsingenieur

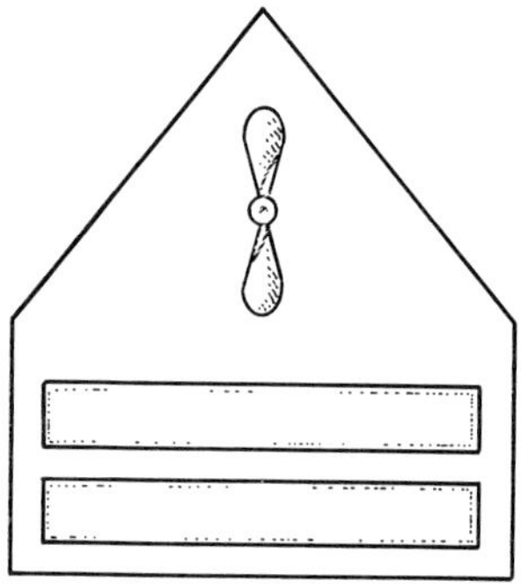

Fl.-Stabsingenieur

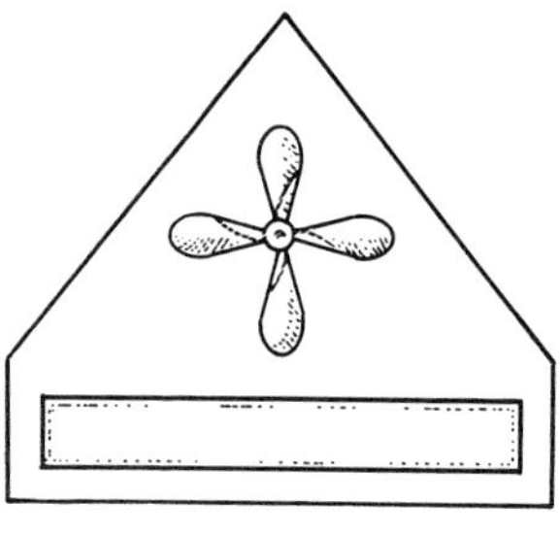

Fl.-Hauptingenieur

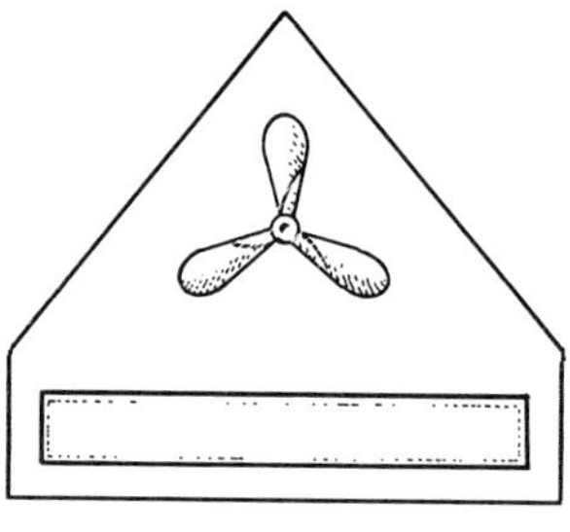

Fl.-Oberingenieur

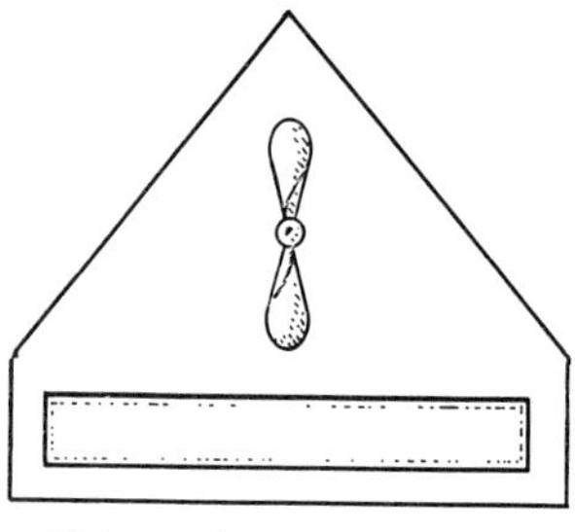

Fl.-Ingenieur

The above propeller and bar ranking system was utilized by engineering officers on both arms of their flying suit (after 4.8.42, on the left arm only). The emblems were in white for grades up to Fl.-Oberstingenieur and in yellow for Fl.-Generalingenieur and Fl.-Generalstabsingenieur. A tan base was used with the light-weight summer flying suit and a blue-grey base for the heavy winter flying suit. Officials also wore the above ranking insignia on their flying suits when on flight status, but with the triangular pip instead of the propeller.

Navigation Corps

Officers in the Navigation Corps (Nautikerkorps der Luftwaffe) wore the same uniform as officers of the Engineer Corps. The Waffenfarben were different however . . . dark green collar patches and shoulder straps with a dark green underlay and a secondary underlay of lemon-yellow (the secondary underlay had been golden-yellow until early 1940). On October 8, 1941, the Navigation Corps Waffenfarben were changed, thusly . . . pink collar patches and shoulder straps with a pink underlay and a secondary underlay of light blue.

Fl.-Stabsnautiker

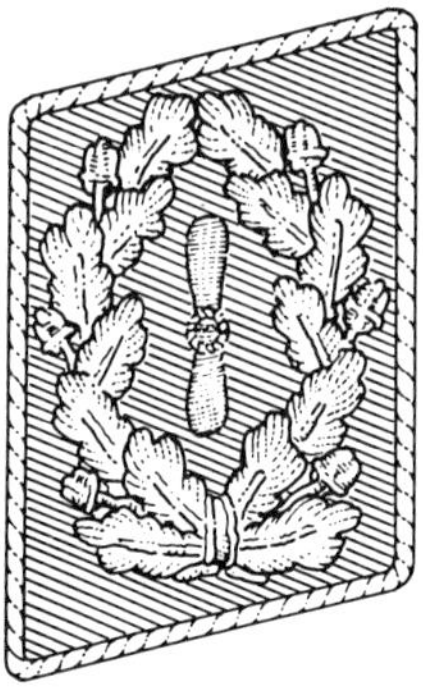

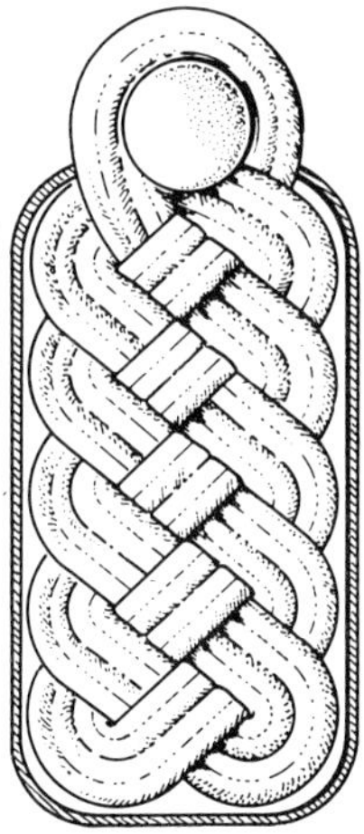

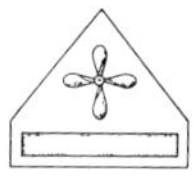

Fl.-Hauptnautiker

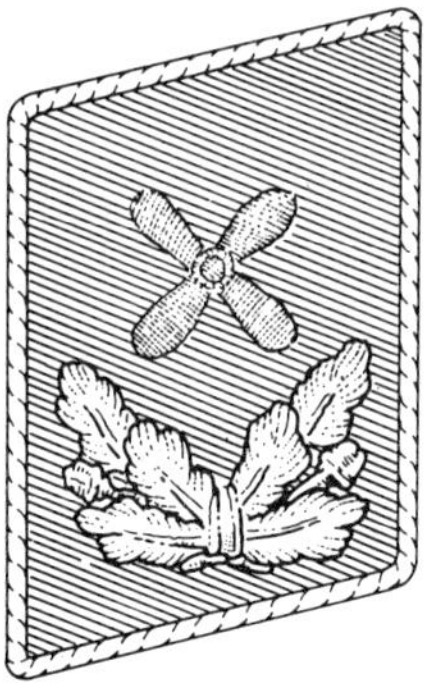

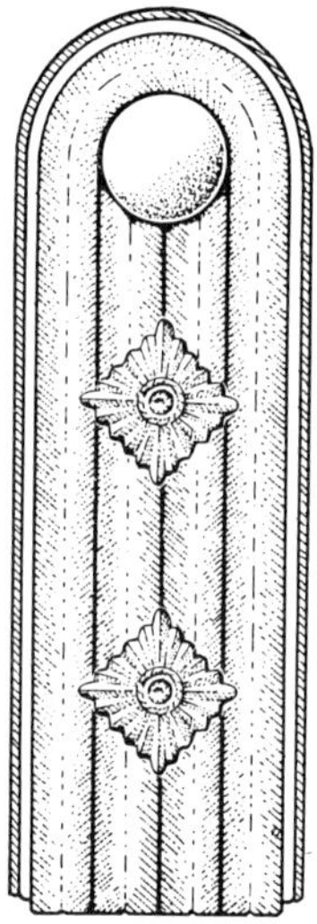

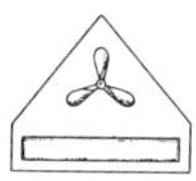

Fl.-Obernautiker

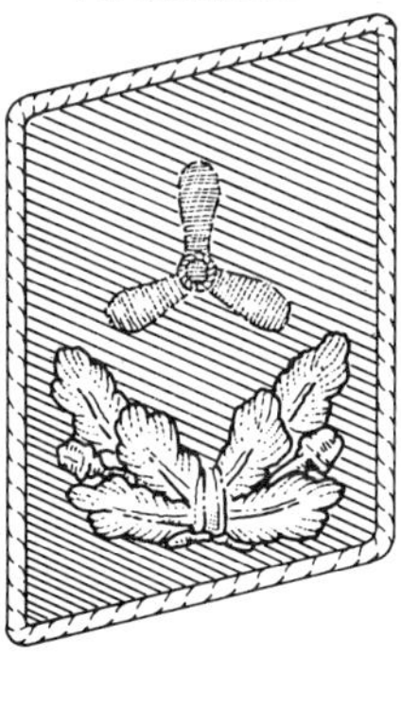

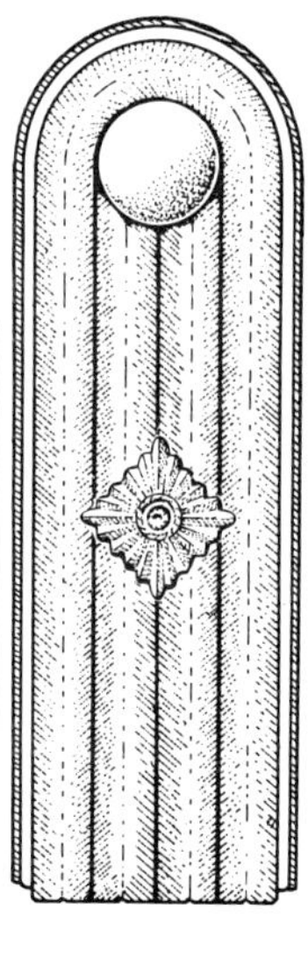

Ordnance NCOs and Officers

Initially, ordnance officers (Offiziere des Waffenwesens) wore the Waffenfarbe of their branch of the Luftwaffe. Shortly after the outbreak of war, however, bright-red became the Waffenfarbe for ordnance personnel. The distinctive Waffenfarbe for reserve ordnance officers had been orange, but was also discontinued during the war. Officers were further idientified by a gold-colored "W" on their shoulder straps.

Ordnance NCOs (Waffenunteroffiziere) wore on their lower left sleeve the following specialty patches:

a. Flying and Signals troops – two crossed rifles

b. Flak Artillery and Regt. General Göring – two crossed cannons (during pre-war years)

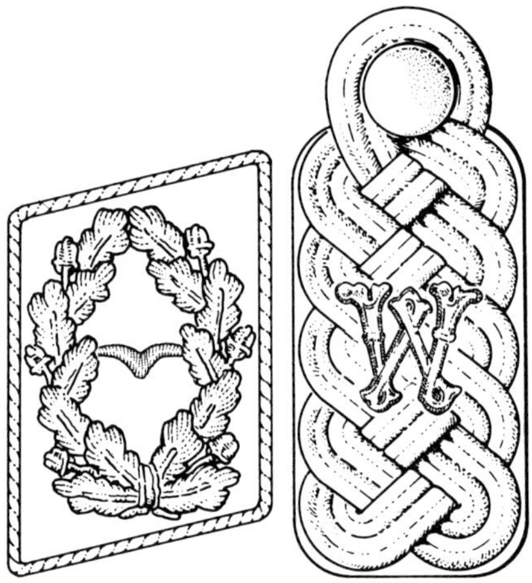

Major (W)

NOTE:

By an order dated 31.3.44, officers (W) became regular Fliegertruppe or Flak officers (the "W" was no longer worn).

Medical Corps

Medical officer candidates (Unterärzte) and officers (Sanitätsoffiziere) wore standard Luftwaffe uniform and rank insignia but with a dark blue Waffenfarbe.[5] The rank designations were, however, as those of medical officers in the Army.

Generaloberstabsarzt (rank insignia as General der Flieger, etc.)
Generalstabsarzt (as Generalleutnant)
Generalarzt (as Generalmajor)

Oberstarzt (as Oberst)
Oberfeldarzt (as Oberstleutnant)
Oberstabsarzt (as Major)
Stabsarzt (as Hauptmann)
Assistenzarzt (as Leutnant)
Unterarzt (Oberfähnrich)
Oberfähnrich im San.Korps
(in the Medical Academy)

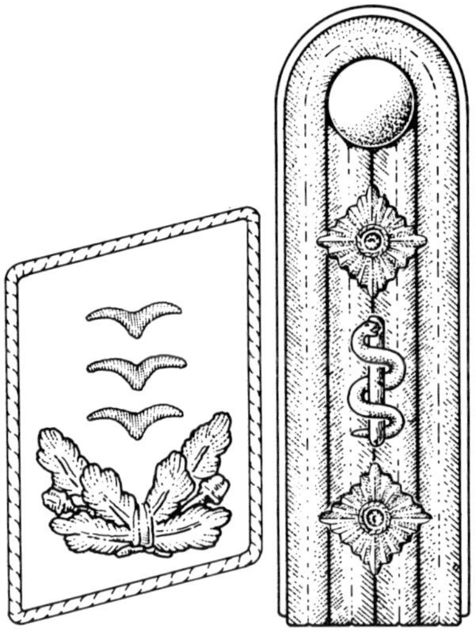

Stabsarzt

An AEsculapius Staff (Äskulapstab) was worn on the shoulder straps of all ranks with the following specifications:

(a) officers – gold-colored metal AEsculapius Staff

(b) officers of general's rank, recalled officers in service and Unterärzte (active and reserve) – aluminum AEsculapius Staff

(c) officers in the Landwehr (reserve officers over 35 years of age) and Unterärzte in the Landwehr – matt-grey oxidized AEsculapius Staff

MEDICAL OFFICER CANDIDATES (UNTERÄRZTE)

The medical officer candidate (Sanitätsoffizieranwärter) wore the uniform of an Oberfeldwebel, with dark blue Waffenfarbe and an aluminum AEsculapius Staff between the two pips on the shoulder straps. The forage cap was piped in aluminum cord and the collar had an aluminum piping (as for officers). A belt in light brown leather was also worn. The peaked cap carried Waffenfarbe piping (not aluminum as for officers) and metal insignia (not embroidered). The medical personnel specialty patch was not permitted to be worn on the lower sleeve until the candidate had passed the first course in medical school. The above uniform was also worn by the Oberfähnrich in the Medical Academy, but with the Gothic "A" on the shoulder straps.

[5] Medical officers of general's rank had white Waffenfarbe and recalled medical officers had orange Waffenfarbe. In 1944, generals' Waffenfarbe was changed to dark blue with the exception of their trouser stripes which remained white (see general officer's Waffenfarbe chart in Waffenfarben section).

LUFTWAFFE MEDICAL ACADEMY

Members of the Luftwaffe Medical Academy wore a Gothic "A" on both shoulder straps, with the following specifications:[6]

(a) all medical officers of the academy – gold-colored "A" and AEsculapius Staff

(b) NCOs with portepee (including Fähnriche and Oberfähnriche) – aluminum "A" and no AEsculapius Staff

(c) NCOs and men (including Fahnenjunker) – dark blue embroidered "A" with light blue piping and no AEsculapius Staff

NOTE:
NCOs and men at the Medical Academy wore the AEsculapius Staff specialty patch on the lower left sleeve after passing their first course (not Oberfähnrich).

FEMALE MEDICAL PERSONNEL

Female medical personnel (Sanitätshelferinnen) were para-military personnel who wore the same uniform as that of "Nachrichtenhelferinnen der Luftwaffe." Rank designations were unique to the medical profession, however, and a medical personnel specialty patch was worn on the lower left sleeve.

CHEMISTS

Chemists (Apotheker) were Wehrmachtbeamte des höheren Dienstes and wore the Wehrmachtbeamten uniform. Teir shoulder straps carried two underlays . . . the basic color being dark green and the inner underlay being apple-green (after 1940, orange was worn as the second color). The collar patches were also dark green as for the Wehrmachtbeamten (with triangular pips). NCOs (Unterapotheker) were of a soldier status and wore the uniform of the Unterärzte without the AEsculapius Staff.

DENTISTS

Dentits (Zahnärzte) had, from the beginning of the war, the status of a medical officer and wore the uniform of the same. no "Z" was worn below the AEsculapius Staff, as for the Army dentist.

[6] Luftwaffen-Verordungsblatt, October 6, 1941. The order was numbered 1361 and dated September 15, 1941.

VETERINARIANS

Veterinary personnel (Veterinärpersonal) were widely used by the German Armed Froces, due to the important role played by horses in military transportation. During the war, a good portion of them were assigned to Luftwaffe Field Divisions. NCOs and men wore the Luftwaffe uniform with crimson Waffenfarbe (collar patches and shoulder strap pipings). NCOs also wore the Farrier specialty patch with light grey piping on the lower right sleeve while other ranks wore the same patch, but without the piping.

Veterinary officers were transferred from the Army and were permitted to retain not only their Army uniform but their Army rank designation.

Specialists

OFFICERS, OFFICIALS AND ENGINEERS (AFTER 7.10.39)

The Sonderführer or specialist was usually a man with civilian specialist qualifications who had been taken into the Wehrmacht and given a rank appropriate to his skill, but above his reserve rank (if any). They were to remain in the Wehrmacht for the duration of the war where they would perform administrative, legal or technical services. Those assigned to the Luftwaffe wore the blue-grey officers' uniform with the following rank insignia alterations:

(a) no wings, triangular pips or propellers were on the officers' collar patches (only the embroidered oakleaves)

(b) a special cord-like shoulder strap was introduced for wear

Sonderführer	Appointment as:
Stellengruppe (R) -	Regimentsführer
Stellengruppe (B) -	Bataillons- or Abteilungsführer
Stellengruppe (K) -	Kompanieführer
Stellengruppe (Z) -	Zugführer
Stellengruppe (O) -	Hauptfeldwebel
Stellengruppe (G) -	Gruppenführer

Sonderführer (R), Wehrmachtbeamter a.Kr., Ingenieur a.Kr. (Rank as Oberst)

The Sonderführer, Wehrmachtbeamte a.Kr. (auf Kriegsdauer – for the war's duration) and Ingenieure a.Kr. in officers rank wore shoulder cords in a smooth silver finish with yellow rayon slides.

Initially, collar patches for Ingenieure and officials had no oakleaf embroidery and only aluminum piping. The illustrated collar patches were, however, utilized during the war. Arm rank insignia, when worn by Sonderführer officers, all officials and Ingenieure des höheren Dienstes on flight suits, was as for active ranks.

Sonderführer (B), Wehrmachtbeamter a.Kr., Ingenieur a.Kr. (Rank as Major)

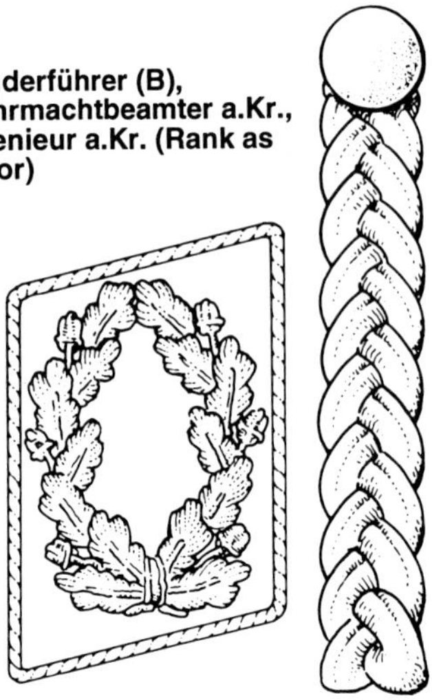

Sonderführer (K), Wehrmachtbeamter a.Kr., Ingenieur a.Kr. (Rank as Hauptmann)

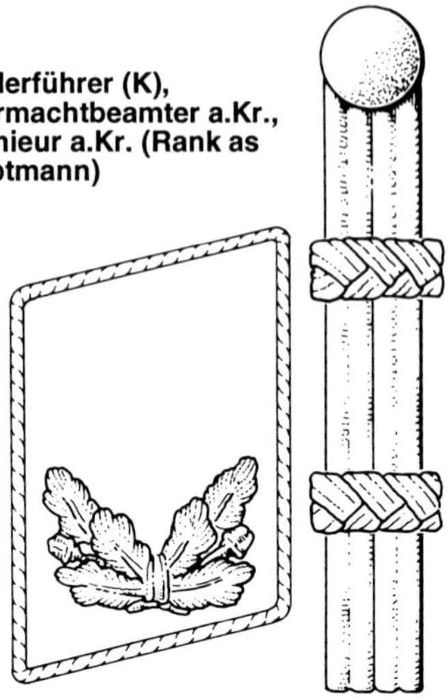

Only Wehrmachtbeamter a.Kr., Ingenieur a.Kr., (Rank as Oberleutnant)

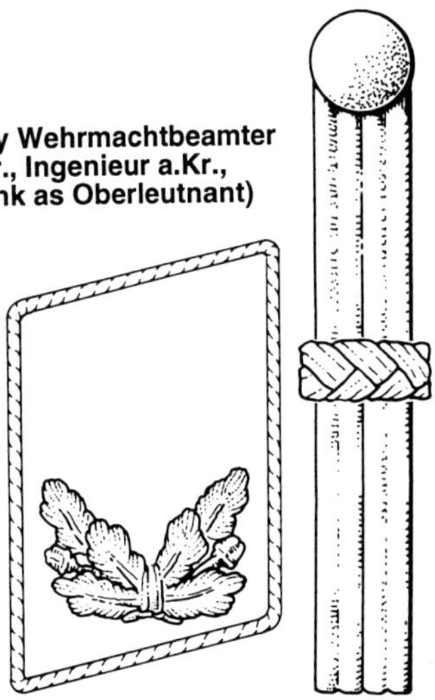

Sonderführer (Z), Wehrmachtbeamter a.Kr., Ingenieur a.Kr. (Rank as Leutnant)

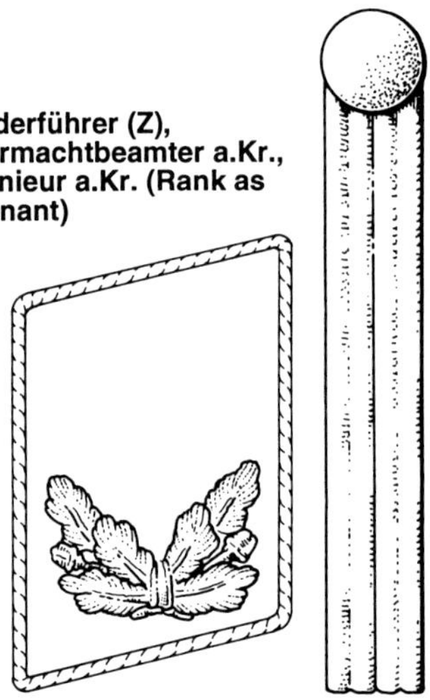

Wehrmachtbeamte a.Kr. in a Feldwebel rank had entwined shoulder cords in dark green with the center cord in aluminum (also one aluminum slide). The collar patches were without insignia.

Wehrmachtbeamte a.Kr. in an Oberfeldwebel rank (Werkmeister, Prüfmeister, Assistenten) had the same dark green, entwined shoulder cords but with two aluminum slides. The collar patches were as those for Assistenten with only the matte-grey embroidered oakleaves and aluminum/dark green/aluminum piping.

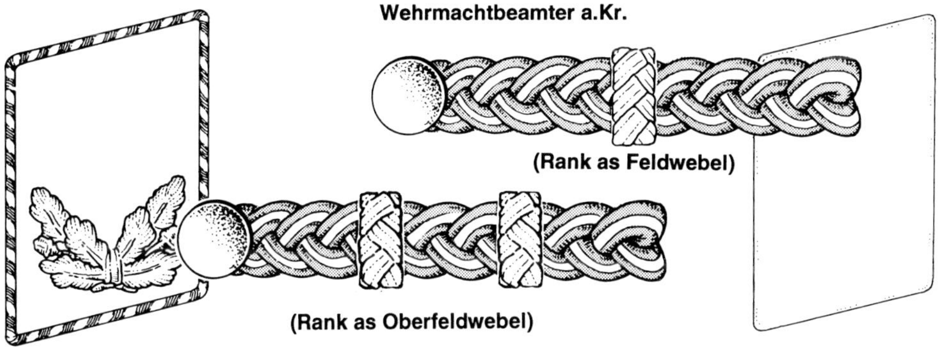

The Gruppenführer and Hauptfeldwebel Sonderführer ranks had their shoulder cords in a bluish/dark green with aluminum slides. The collar patches were without any insignia and the collar was generally piped in the appropriate Waffenfarbe. As was true for all other Sonderführer ranks, the Waffenfarbe worn was as that of the branch to which they were assigned.

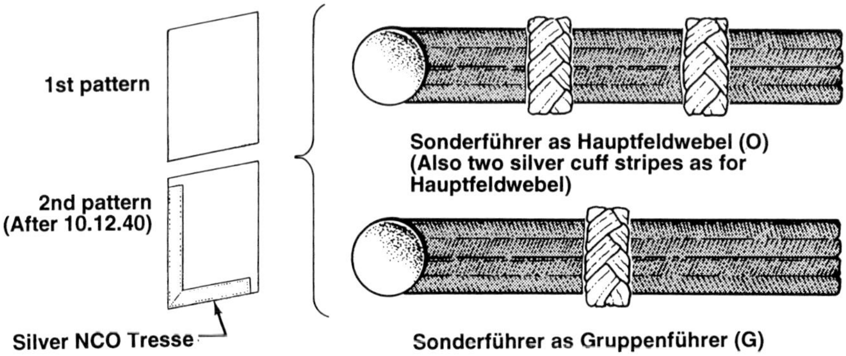

Reserve and Landwehr Officers

The reserve* and Landwehr* officer wore the uniform of the Luftwaffe with the following variations:[7]

(a) The collar patches were in the Waffenfarbe of their servcie branch with a light blue border and aluminum piping.

(b) Shoulder straps had a double underlay . . . the base one being in the Waffenfarbe of their service branch and the inner underlay being in light blue.

(c) Reserve officers initially wore a Reich cockade with a Landwehr cross center on their peaked cap.

(d) Reserve officers wore the gold metal number of their unit (reserve medical officers wore a gold-colored AEsculapius Staff) on their shoulder straps.

(e) Landwehr officers wore Roman numerals in oxidized metal on their shoulder straps which indicated the Luftgaukommando (Air District Command), and gold pips.

(f) Wehrmachtbeamte des Beurlaubtenstandes** (officials) wore the uniform as above but with a light blue border on dark green collar patches piped with gold or aluminum cord. The shoulder straps had a double underlay . . . the base one being in dark green and the inner underlay being in light blue.[8]

(g) Reserve and Landwehr officers in the Engineer Corps wore the uniform as above but with a light blue border on pink collar patches piped with aluminum cord. The shoulder straps had a double underlay . . . the base one being in pink and the inner underlay being in light blue, with gold pips.

* Reserve Officers (Offiziere d.R.) – selected for commissions at the time they had completed their two years service as conscripts or their twelve years service as professional soldiers.
* Landwehr Officers (Offiziere d.L.) – a qualification formerly used for reserve officers over 35 years of age.
** Beurlaubtenstand (Offiziere d.B.) – a collective term for reserve and Landwehr officers.

Landwehr cross cockade (worn only for a short time after 1935)

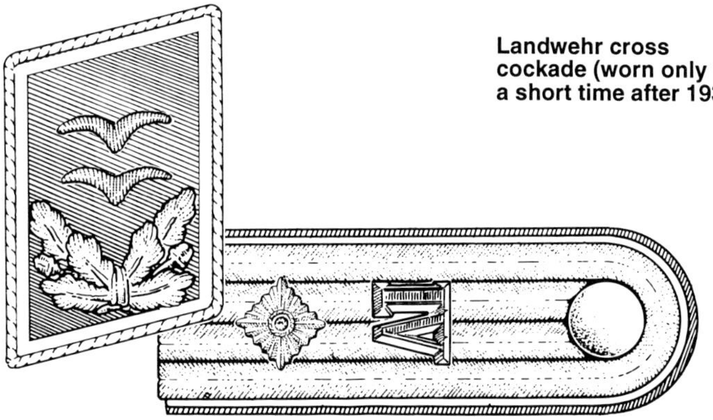

Landwehr-Oberleutnant from the IV Luftgau

[7] E.J. Graf von Westarp, *Taschenkalender für die Luftwaffe* (Grimmen in Pommern: Verlag Alfred Waberg, 1942), p.1169.

[8] Ibid, p.1192, footnote 2. Wehrmachtbeamte des Beurlaubtenstandes now wore, for the duration of the war, the same insignia as active Wehrmachtbeamte with their specialty Waffenfarbe as the inner underlay of the shoulder straps.

Officers, etc., z.V.*

Former active officers and Offiziere des Beurlaubtenstandes of the old Wehrmacht who were not in the Offizierkorps des Beurlaubtenstandes of the new Wehrmacht (after 1935) may wear the uniform of the reserve officer, official or engineer with appropriate Waffenfarbe, his rank at retirement and no regimental numbers, etc. on the shoulder straps. Two Tressen were added to the bottom of the shoulder straps after being mustered out of the service to indicate an individual was retired. This uniform could only be worn in service, to and from service and in the officer's home.

* Officers, etc., z.V. (zur Verfügung) – Regular or reserve officers who have been technically retired upon completion of their term of service, but who have been retained in service or recalled for a new assignment (two aluminum Tressen).

NOTE:

Retired personnel carried one Tresse (aluminum) and generals a.D. had the underlay of their shoulder straps in the Waffenfarbe of their old unit plus collar patches, etc. in white.

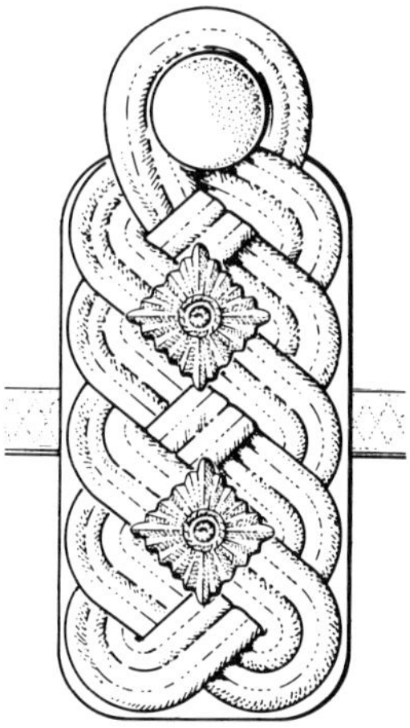

Retired (Oberst a.D.)

Officer Candidates

The Fähnrich (officer candidate) wore the uniform of an Unterfeldwebel but with an aluminum portepee for the model 84/98 bayonet in place of the bayonet knot. The Waffenfarbe was as that of his branch and the numbers and letters on the shoulder straps were in an aluminum-colored metal. The flyer's sword could be worn without the sword knot, but with the portepee. With the Walking-Out dress, the flyer's dagger could be worn with portepee.

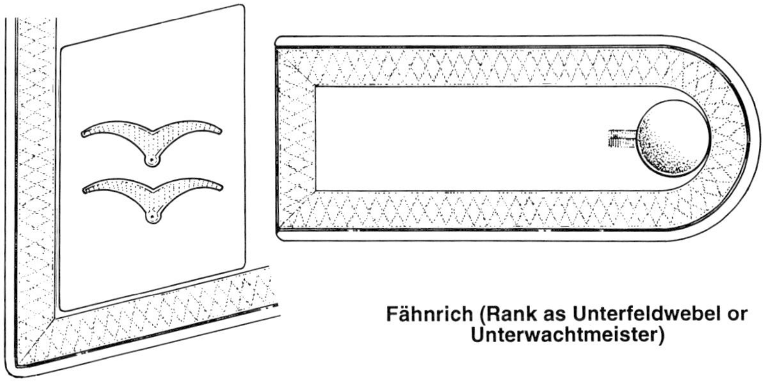

Fähnrich (Rank as Unterfeldwebel or Unterwachtmeister)

The Fahnenjunker (appointed officer candidate) wore the uniform of his rank with a promotion to Gefreiter or Unteroffizier, and with the Waffenfarbe of his own branch of service. He also wore the bayonet knot for NCOs. Around the base of his shoulder straps were worn aluminum cords of the same dimension as that around the officer's collar.

Fahnenjunker-Gefreiter

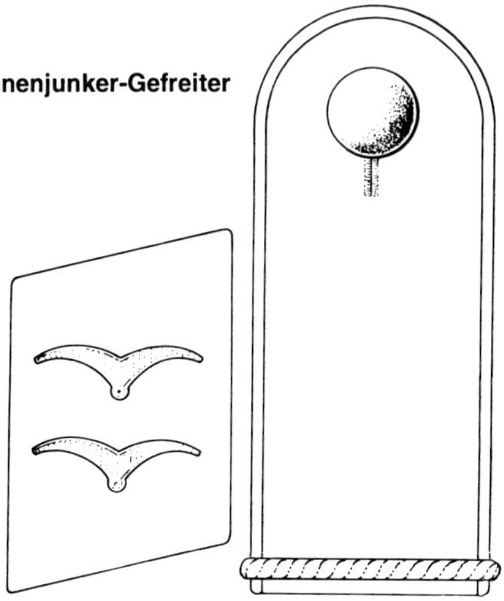

The Oberfähnrich (advanced officer candidate) wore the uniform of an Oberfeldwebel but with alterations. Aluminum cap cords and embroidered insignia were worn on the peaked cap. The peaked cap was, however, piped in the candidate's appropriate Waffenfarbe. The forage cap was for officers. No NCO Tresse was worn around the collar of the flying blouse or service tunic, but an aluminum cord for officers was. In place of the dark brown belt for NCOs and men, the 6cm wide, light brown, double claw belt and buckle plus cross-over strap for officers was worn.* Unterärzte were as above but with aluminum-colored AEsculapius Staffs between the two pips on the shoulder straps.**

Oberfähnrich

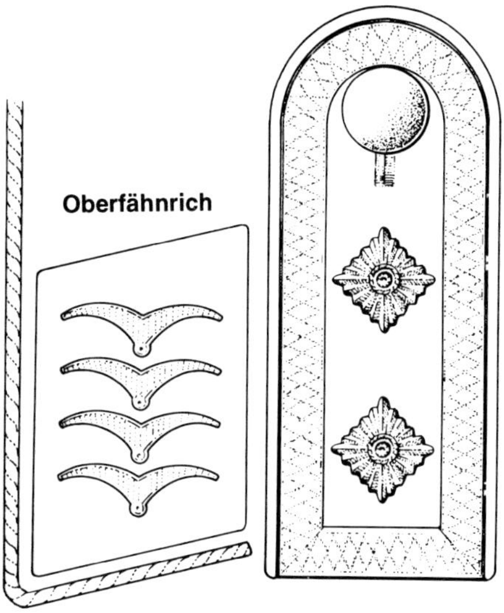

Oberfähnrich
Herbert Bertholdt

NCO Candidates

The Unteroffizieranwärter (NCO Candidate) wore the uniform of his rank and service but with a 1cm aluminum Tresse positioned on the base of his shoulder straps.

The Unteroffizierewärter (Ergänzungs (retired then recalled) NCO candidate) wore on the lower portion of his shoulder straps, a patterned braid, 0.8cm wide and in green.

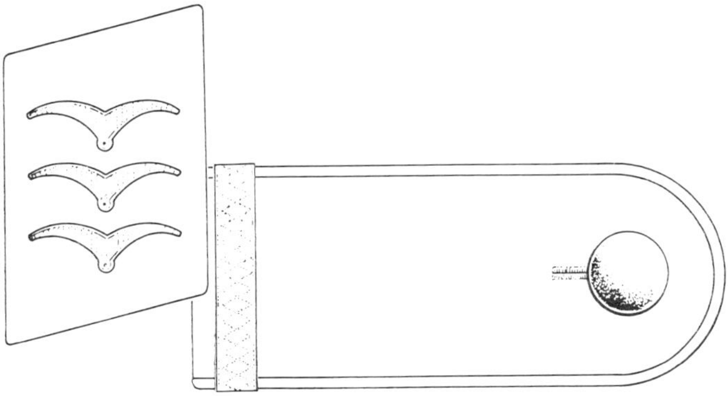

Unteroffizieranwärter (here an Obergefreiter)

NCO SCHOOLS OF THE LUFTWAFFE

The uniform for students of Luftwaffe NCO schools 1, 2, 3, and 4 are as for the Fliegertruppe. All students wore on the lower right sleeve of the tunic (approximately 1cm above the cuff and in the same position for the flying blouse) a "US" (Unteroffizierschüler) specialty patch in matte-grey thread on a blue-grey base. The school cadres wore the uniform of their parent unit and "US" on their shoulder straps (officers – gold, NCOs with portepee – aluminum, NCOs and men – embroidered in their own Waffenfarbe).

NOTE:

Teachers in technical schools were considered Luftwaffen-Wehrmachtbeamte and wore the official's uniform. Excluded are teachers in the testing service.

* The wearing of the cross-over strap was discontinued shortly after the war began.

** The Fahnjunker (San.) and Fähnrich (San.) did not wear the AEsculopius Staff on their shoulder straps.

Waffenfarben

White – Generals (including medical officers of general's rank until 1940), also Regiment General Göring (later Division Hermann Göring)
Golden-Yellow – Flying Troops (also other staffs, units and branches)
Golden-Brown – Signals Troops
Bright-Red – Flak Artillery, Ordnance Officers
Crimson – Officers of the General Staff
Pink – Engineer Corps
Bordeaux Red – Court Martial Officials
Wine Red – Judicial Officials
Orange Red – Recalled Officers (during the war, the uniform for active officers was worn)
Dark Blue – Medical Officers and Personnel
Light Green – Air Traffic Control Officials
Black – Officers, NCOs and men in the RLM (Reichsluftfahrtministerium) until June 30, 1939.[9] RLM personnel were now to wear the Waffenfarben of their parent unit and those who had no distinctive colors before entering RLM service, wore golden-yellow. By an order dated December 22, 1939, Luftwaffe Construction Engineers were now to wear black.[10]
Dark Green – Wehrmachtbeamte

Additional Colors Worn:
Bright-Red – NCOs and men in Regiment General Göring (Flak Abteilung) as bordering to their white collar patches (up to 1942 when it became brigade strength).
Rifle Green – NCOs and men in Regiment General Göring (except those from the Flak Abteilung) as bordering to their white collar patches (up to 1942 when it became brigade strength).
Light Blue – Officers, Officials and Engineers (reserve and Landwehr) as bordering to their collar patches and the inner underlay to their shoulder straps. This color configuration was not worn during the war.

NOTES:
1. See the Corps of Administration Officials section for the complex color combinations worn by various administrative officials.
2. See Division Hermann Göring and Field Divisions sections for additional Waffenfarben listings.

[9] Luftwaffen-Verordungsblatt, July 10, 1939. The order was numbered 638 and dated June 30, 1939.

[10] Luftwaffen-Verordungsblatt, January 2, 1940. The order was numbered 4 and dated December 22, 1939.

WAFFENFARBEN FOR GENERAL OFFICERS AND OFFICIALS OF GENERAL OFFICER RANKS

Corps or Branch	Collar Patches		Shoulder Straps			Pants Stripes	Great-Coat Lapels	Remarks
	Color	Oak-leaves and	Underlay		Insignia			
			1st Color	2nd Color (inner)				
1. Officers:								
(a) Offizierkorps (also retired but not including 1b)	white	gold wings	white	-	-	white	white	
(b) Retired Generals (with uniform of a unit)	white	gold wings	yellow, red or brown	-	- silver no. -	white white white	white white white	
(c) Sanitätsoffizier-korps	white	gold wings	white	-	silver ⚕	white	white	
2. Officials:								
(a) all branches except 2b and 2c	dark green	gold pips	dark green	red	-	dark green	dark green	till 16.2.40
(b) Reichskriegs-gericht	dark green	gold pips	dark green	bordeaux red	-	dark green	dark green	till 5.12.39
(c) Ingenieurkorps	pink	gold pro-peller	pink	-	-	pink	pink	
1. Officers:								
(a) same as A1a								
(b) same as A1b								
(c) same as A1c								
2. Officials:								
(a) all branches except 2b, 2c, and 2d	dark green	gold pips	dark green	white	-	white	dark green	after 16.2.40*
(b) Reichskriegs-gericht	bordeaux red	gold pips	dark green	bordeaux red	-	bordeaux red	bordeaux red	after 5.12.40*
(c) Justizverwalt-ung	wine red	gold pips	dark green	wine red	-	wine red	wine red	after 5.12.40*
(d) Ingenieurkorps same as A2c								

*Allg. Verwaltungsdienst, Reichskriegsgericht and Justizverwaltung became "Offiziere im Truppensonderdienst" by an order dated May 9, 1944.

1944/1945

Corps or Branch	Collar Patches: Color	Collar Patches: Oak-leaves and	Shoulder Straps: Underlay 1st Color	Shoulder Straps: Underlay 2nd Color (inner)	Shoulder Straps: Insignia	Pants Stripes	Great-Coat Lapels	Remarks
1. Officers								
(a) same as A1a								
(b) same as A1b								
(c) Sanitätsoffizier-korps	dark blue	gold wings	dark blue	-	silver	white	dark blue	after 1.5.44 (before as B1c)
(d) Truppensonder-dienst								
- Verwaltungs-dienst	light blue	gold wings	light blue	-	silver	white	light blue	after 1.5.44 (before as B2a)
- Richter	wine red	gold wings	wine red	-	silver	white	wine red	after 1.5.44 (before as B2b and B2c)
2. Officials:								
(a) Allgemeine Verwaltung	dark green	gold pips	dark green	red	-	dark green	dark green	after 1.5.44 (before as B2a)
(b) Bautechnische Verwaltung	dark green till 19.8.44 (black after 19.8.44)	gold pips	dark green	red till 19.8.44 (black after 19.8.44)	-	dark green till 19.8.44 (black after 19.8.44)	**dark green till 19.8.44 (black after 19.8.44)**	after 1.5.44 (before as B2a)
(c) Wetterdienst	dark green	gold pips	dark green	blue-grey	-	dark green	dark green	after 1.5.44 (before as B2a)
(d) Deutscher See-wetterdienst	dark green	gold pips	dark green	blue-grey	-	dark green	dark green	after 1.5.44 (before as B2a)
(e) Fachschuldienst	dark green	gold pips	dark green	orange	-	dark green	dark green	after 1.5.44 (before as B2a)
(f) Bibliotheksdienst	dark green	gold pips	dark green	orange	-	dark green	dark green	after 1.5.44 (before as B2a)
(g) Medizinaldienst	dark green	gold pips	dark green	orange	-	dark green	dark green	after 1.5.44 (before as B2a)
(h) Apotheker	dark green	gold pips	dark green	orange	-	dark green	dark green	after 1.5.44 (before as B2a)
(i) Chemiker	dark green	gold pips	dark green	orange	-	dark green	dark green	after 1.5.44 (before as B2a)
(j) Kriegswissen-schaftliche Abteilung	dark green	gold pips	dark green	orange	-	dark green	dark green	after 1.5.44 (before as B2a)
(k) Reichsflug-sicherungs-dienst	dark green	gold pips	dark green	dark brown	-	dark green	dark green	after 1.5.44 (before as B2a)
(l) Ingenieurkorps same as B2d								

2

4

8

9

12

Additional recognition for special services, members of schools, etc., was made by the adding of letter or number combinations, plus identifying insignia onto the uniform shoulder straps. Listed below are the known variations, their service or school and proper Waffenfarbe.

1. A – Luftwaffe Medical Academy (dark blue)
2. AL – Reconnaissance Training Group (golden-yellow)
3. F – Fortress Flak Artillery Battalion, over the Battalion no. (red)
4. FAS – Flak Artillery School (red)
5. FS – War Ordnance School (golden-yellow)
6. GFP – Secret Field Police (inner underlay over dark green is wine-red)
7. HS – Advanced Luftwaffe School (golden-yellow)
8. KA – Air War Academy (golden-yellow)
9. KRS – Transport School (golden-yellow)
10. KS – Air War School (golden-yellow)
11. L – Training Unit Cadres (Waffenfarbe of the Cadre's parent unit)
12. LS – Reich School for Air Controllers (light green)
    – Luftschützenheiten (as black with white piping)
13. LW – Luftwaffe Garrison Battalion Cadre in Berlin (golden-yellow)
14. NS – Air Signals School (golden-brown)
15. OL – Signals Battalion of the Oberkommando der Luftwaffe (golden-brown)
16. RL – National Institute for Air Defense (red)
17. RLM – Reich Air Ministry (Waffenfarbe of their parent unit) plus:*
    (a) Flight Readiness of the RLM (golden-yellow)
    (b) Main Photo Department of the RLM (golden-yellow)
    (c) Main Moving Film Department of the RLM (golden-yellow)
    (d) Information Service Department of the RLM (golden-yellow)
18. SL – Training Staffel (Sea) (golden-yellow)
19. SS – Air Sports School (golden-yellow)
20. US – NCO School Cadre (golden-yellow)
21. W – Ordnance Officers (red)
22. WS – Air Ordnance School (golden-yellow)
23. WS – Flak Ordnance School (red)
24. Block Numbers – Flak Artillery Regiments (red)
25. Roman Numerals – Landwehr officers (indicates Air District Command)
    – Wehrersatzdienststellen (yellow)
26. Lyre – Music Corps (Waffenfarbe of their parent unit), and Band Masters
27. AEsculapius Staff – Medical Corps (dark blue)
28. Judicial Sword – Judicial Officers (inner underlay over dark green is wine red – after 1.5.44 only)
29. Administrative Insignia – Administrative Officers (inner underlay over red – light blue after 1.5.44 only)

14

15

17

18

19

The Luftwaffe officer is from the RLM and wears the red helmet and cap band, indicating the "Rote Partei" in maneuvers. The band was reversible to yellow which indicated maneuver command and neutral or dead persons in maneuvers (1936 winter maneuvers of Gruppenkommando 2 and I. & IV. Armeekorps). Note "Sepp" Dietrich and von Ribbentrop on the left.

Uniforms and Headgear

Breast Eagles

This air force version of the national emblem was worn over the right breast pocket with the swastika overlapping the pocket flap. It was machined in matte-grey thread for NCOs and men, but could be hand-embroidered on non-issue (private purchase) tunics. The breast eagle for officer ranks was in hand-embroidered aluminum thread and for general ranks in hand-embrodered gold thread.

Machine-embroidered eagle for NCOs and men

Machine-embroidered eagle in light aluminum thread for non-issue tunics

Hand-embroidered eagles for officers (note drooped tail of eagle above, indicating an early manufacture (1935-1936))

Uniforms

By Hitler's decree of February 26, 1935, the Luftwaffe was to be offically the third branch of the Wehrmacht as of March 1, 1935. The new Luftwaffe was faced with the problem of uniforms . . . they wanted a distinctive uniform from those of the other two branches of the Wehrmacht (Heer and Kriegsmarine), and also wanted a clear separation in dress of military and civilian flyers.

Preceding the March 1 date, all those in secret training for the Luftwaffe wore the basic Deutscher-Luftsport-Verband uniform with minor variations. After careful examination of production costs of an entirely new uniform, it was decided to utilize the basic grey-blue DLV uniform with newly designed Luftwaffe insignia. Among these were the Luftwaffe national emblem (a flying eagle clasping a swastika), a new and extensive use of Waffenfarben, some changes in the DLV collar rank insignia and the adoption of the Army's shoulder strap ranking system.

With the passage of time, a number of various uniforms evolved in cut and design which were designated to cover any occasion Luftwaffe personnel would encounter. Among these were:

(a) Flying Service Uniform for flying personnel
(b) Field Dress
(c) Service Dress
(d) Guard Uniform
(e) Undress Uniform for officers and NCOs which lead (on duty)
(f) Reporting Uniform
(g) Parade Dress
(h) Walking-Out Dress
(i) Informal Full Dress (day) for officers
(j) Formal Full Dress (day) for officers
(k) Informal Full Dress (evening) for officers
(l) Formal Full Dress (evening) for officers
(m) Informal Full Dress for NCOs and men
(n) Formal Full Dress for NCOs and men
(o) Summer Uniform for officers
(p) Sports kit

Service Tunic

The four-pocket Service Tunic, whose design had been borrowed from the DLV in 1935, was of a blue-grey material with matte-aluminum buttons. Standard Luftwaffe collar and shoulder rank insignia were worn with this tunic as well as the Luftwaffe national emblem over the right breast pocket. All NCOs and men had the collar piped in their appropriate Waffenfarbe,[1] officers and some NCOs with portepee (only officer candidates) had a 1cm wide aluminum piping and generals had a 1cm gold colored piping. When the Service Tunic was worn with a blue shirt and black tie, it served as the tunic for:

(a) Parade Dress
(b) Service Dress for officers and NCOs with porteppe
(c) Undress Uniform for officers
(d) Reporting Dress
(e) Walking-Out Dress (also worn with white shirt)
(f) Flying Service Uniform for all flying personnel

With white shirt and black tie:

(g) Formal and Informal Full Dress for NCOs and men
(h) Formal and Informal Full Dress (day) for officers

NOTE:
The tunic for NCOs and men was in wool and that for officers was usually in garbardine.

Luftwaffe enlisted men in parade dress.

[1] This practice was discontinued on the Flying Blouse by an order dated March 20, 1940.

Uniform Tunic

The blue-grey, four pocket Uniform Tunic was similar to the Service Tunic with the exceptions that it had an extra button plus a special collar cut allowing it to be worn open or closed. It was designed, embodying features of the Service Tunic and Flying Blouse, to eventually replace both of them. It was not intorduced for wear by Luftwaffe personnel until November 1938, allowing the old style Service Tunic to be worn for a number of war years.

Generaloberst Dessloch greeting German and Rumanian fighter pilots in July 1944 (note collar being worn open).

Flying Blouse

Officer's (Oberleutnant) flying blouse with side pockets

The blue-grey flying blouse was introduced for wear shortly after the Luftwaffe's inception. It was shorter than the service tunic, had one row of hidden plastic buttons and initially had no side pockets.[2] Rank insignia and the Luftwaffe breast eagle were as on the service tunic.[3]

The officers' model was as above, but without belt suppoters, usually in gabardine, and with a dagger hanger sewn inside the blouse. The flying blouse was worn with open or closed collar such as for:

(a) flying service uniform for NCOs and men
(b) field dress for NCOs and men
(c) service dress for NCOs without portepee and men

Or with blue shirt and black tie for:
(a) flying service dress and field dress for officers
(b) undress uniform for NCOs on leading duty
(c) undress uniform for officers

Note lack of breast eagles and side pockets on pre-1940 flying blouse above.

Generalleutnant Richard Heidrich, commander of the 1. Fallschirmjäger Division.

[2] Luftwaffen-Verordnungsblatt, December 2, 1940, order #1509. By request of many troops, the up-till-now pocketless flying blouse was to have side pockets. It was also ordered that old-style blouses were not to be altered, but worn until discardable. The order was dated November 19, 1940. Officers always had side pockets. After above date, other ranks had side pockets with flaps and buttons.

[3] Luftwaffen-Verordnungsblatt, October 14, 1940, Order #1290. By an order dated October 1, 1940, NCOs and men were now permitted to wear the Luftwaffe eagle on the right breast of their flying blouse. From this date, individuals were to acquire eagles for their blouses and all supplies of flying blouses issued were to have the eagle positioned in the same approximate position as on the service tunic.

Special Tunic for Generals

Generals (including medical officers of Generals' rank) as well as officials (Wehrmachtbeamte) and engineers (Ingenieure) of generals rank wore a special tunic (kleiner Rock für Generale) with Walking-Out Dress, Undress and Informal Full-Dress uniforms. Its basic color and insignia was as that of the Service Tunic. The wide lapels, piping above cuffs and down the front of the tunic were in their appropriate Waffenfarbe: i.e. Generals –white (doeskin), officals – dark green, and engineers – pink.

When worn with a Work Uniform (while on duty in an orderly room), the above style tunic could be worn but without Waffenfarbe colored lapels or piping and without collar patches. Under the flap of the left pocket a 7cm long hanger for a sword or dagger was sewn.

NOTE:
No new tunics of this style were made available during the war. It could, however, be worn during the war years if already possessed by officers of general's rank.

Generalfeldmarschall Göring is greeted on his birthday (January 12, 1939) by generals of the Luftwaffe.

Formal Full-Dress (Evening) for Officers

The mess jacket and straight pants were of a blue-grey material and worn with both Informal and Formal Evening Full-Dress.[4] The insignia worn was an aluminum hand-embroidered Luftwaffe eagle (gold for generals) on the right breast and standard Luftwaffe shoulder straps.[5] The aluminum colored buttons were as those on the Service Tunic (gold for generals). The jacket was worn with a stiff white shirt with peaked collar and mother-of-pearl studs.

The matte-aluminum aiguillette for adjutants was not worn with this dress uniform. All officers wore a bright aluminum or gold aiguillette.

On each leg of the straight pants, worn with the Informal Evening Full-Dress:

(a) by generals, was a white piping running the length of the pants, bordered by two 0.4cm wide blue-grey stripes and then two 4cm wide white stripes.

(b) by officials (Wehrmachtbeamte), with the rank of general –dark green piping and stripes

(c) by engineers (Ingenieurkorps), with the rank of general –pink piping and stripes

(d) by general staff officers – crimson piping and stripes

When general officer ranks wore the Formal Evening Full-Dress, the straight pants were with a 4cm wide gold braid running the length of the pants on both legs. Other officer ranks wore a 4cm wide aluminum braid on both legs.

General Friedrich Christiansen in a Formal Evening Full-Dress

Informal Evening Full-Dress:
Blue-grey jacket and straight pants without stripes, blue waist coat (vest), white shirt, black tie and ribbon bar
Formal Evening Full-Dress:
Blue-grey jacket and straight pants with stripes, white waist coat (vest), white shirt, white tie, aiguillette and medal bar

[4] The naval mess jacket was instrumental in influencing the design of this uniform.

[5] A 3mm wide aluminum piping was also around the collar of the jacket (gold for generals).

Summer Tunic for Officers

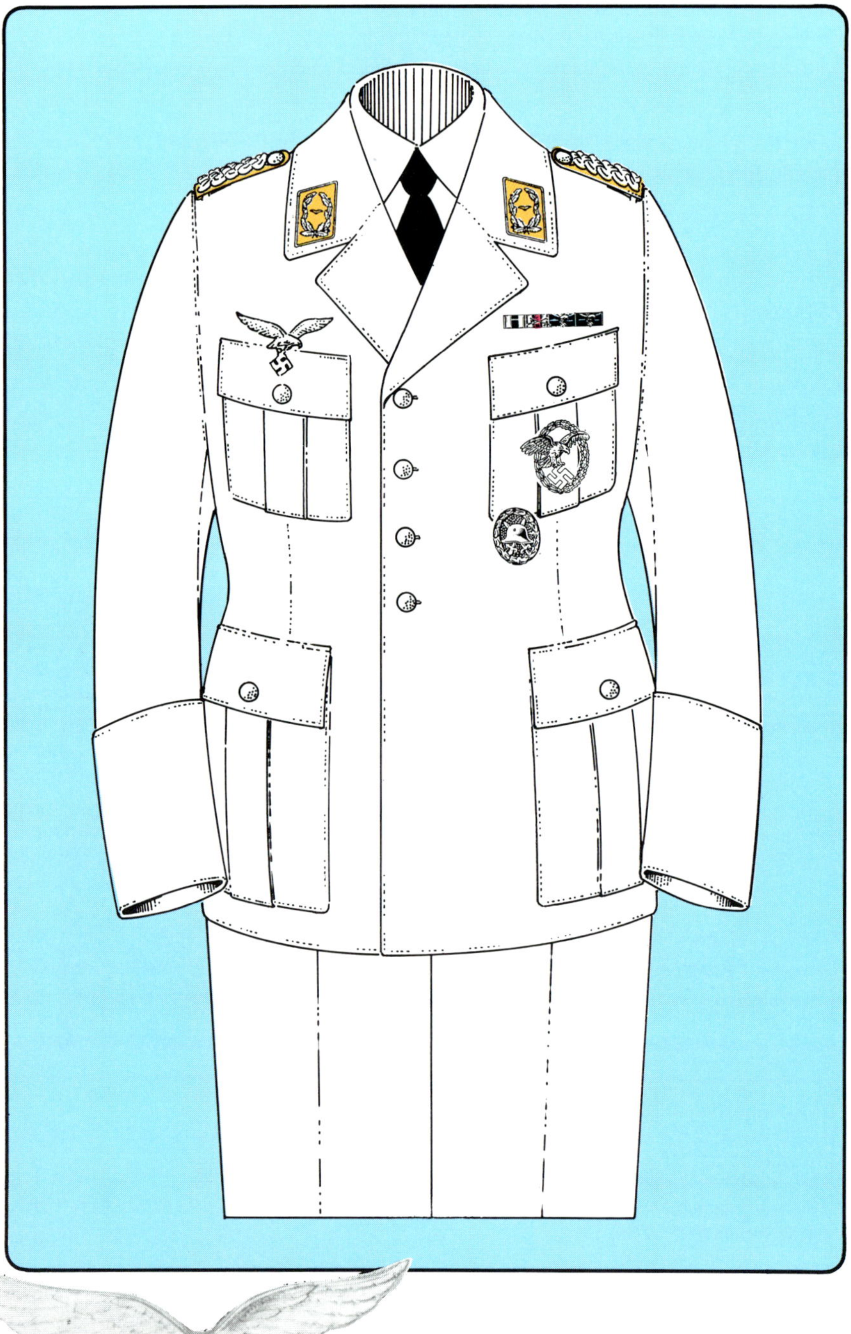

Metal pin-on breast eagle

This tunic was usually in a white gabardine or linen material, the same cut as the Service Tunic but without lining and had removable aluminum buttons (gold for generals). The rank insignia was also removeable. The Luftwaffe breast eagle was in an aluminum colored metal (gold for generals) and pinned above the right breast pocket, with the swastika overlapping the flap of the pocket.[6] The straight pants were also in white gabardine or linen with aluminum buttons (gold for generals)

This tunic was authorized for wear from April 1 to September 30 by officers, officials (Wehrmachtbeamte) of officers' rank and officers of the two corps (Ingenieurkorps and Nautikerkorps). It could be worn with the Walking-Out Dress (blue or white shirt), Undress (blue shirt), Informal or Formal Full Dress (day) (white shirt), and while on duty in the orderly room.

When inside a barracks, on an air base, as well as in an office, an officer could wear the Summer Tunic without collar patches. When worn with the Walking-Out Dress (Ausgehanzug) or Formal and Informal Full-Dress (Gesellschaftsanzug), however, collar patches were required.[7]

Note positioning of collar patches on the summer tunic at right.

[6] No aluminum cord for officers (gold for generals) was worn around the edge of the collar on this tunic.

[7] Luftwaffen-Verordnungsblatt, May 17, 1937. The order was numbered 614 and dated May 4, 1937.

Tropical Uniform

Variation-cut Italian tropical tunic which saw much service in the Mediterranean area.

The tan Tropical Uniform was authorized for wear between the dates of May 1 and September 30, within the recognized borders of Germany. In actual tropical areas such as Sicily and Africa, this uniform was worn year round. The uniform consisted of a forage cap, tunic, long or bloused pants, shirt, tie and lace-up boots. The wearing of the tropical helmet, short pants and shirt without tunic was forbidden in the home area.

The uniform parts were in yellow tan material.[8] The tunic buttons were of pebbled metal and also in tan. The Luftwaffe breast eagle was white on a tan base. Many officers, however, utilized a pin-on, aluminum-colored, metal eagle (gold for generals) as that worn on the white Summer Tunic. The shoulder straps were detachable and standard blue-grey issue or tan. Collar patches were not normally worn with the Tropical Uniform. Chevroned ranks wore their insignia on the upper left arm, either a standard Luftwaffe issue on a blue-grey base or copper-brown Tresse on a tan base.

Generalfeldmarschall Rommel and Generalmajor Fröhlich in Africa.

A tailor's finishing touches to a standard-cut Tropical tunic.

Between the dates of May 1 and September 30, the Tropical Uniform was also worn in Italy (south of the Naples-Foggia line), on Sicily and Sardinia, in Greece and on Crete. When Luftwaffe personnel were assigned to these areas, they were issued the following tropical clothing in addition to their blue-grey uniform:[9]

1 – forage cap
1 – tunic
1 – pair of long or bloused pants
1 – pair of short pants
2 – shirts with long sleeves
2 – shirts with short sleeves
3 – pair of underpants
2 – ties (khaki brown)

[8] It should be noted that the tropical uniform color for the Luftwaffe was quite different from that of the Army. While the Luftwaffe utilized a tan with an orange or yellow tinge, the Army uniform was predominently khaki or of an olive shade.

[9] Luftwaffen-Verordnungsblatt, March 2, 1942. The order is numbered 492 and dated February 14, 1942.

The first Luftwaffe personnel in Africa utilized available stocks of army tropical clothing (Note army breast eagle, pith helmet and Luftwaffe collar patches).

The tropical lace-up boot

Young Luftwaffe paratroopers in tropical shirts and bloused pants. Note detachable shoulder straps and ankle buckles on bloused pants (top of photo).

From October 1 to April 30, troops on duty and on furlough in the Luftwaffe administrative areas of Naples, Luftgau South East and the South Greece Command in Athens, wore the blue-grey uniform. During this designated time, only the blue-grey uniform could be worn within the recognized borders of Germany.[10]

Short pants and shirts with long or short sleeves were permitted to be worn only on air bases, at flak installations, while on duty with a signals unit, in the vicinity of an individual's barracks or while marching in formation as a unit. All other situations required the tunic, or bloused pants and a shirt with tie.

Clothing inspection . . . in short pants!

In early 1944, the issuance of tropical clothing to Luftwaffe personnel was as follows:[11]

1 – soft peaked cap with neck protector or (1) forage cap
1 – tunic
1 – pair of bloused pants
2 – shirts with long sleeves
2 – undershirts
1 – tie
2 – pair of underpants
2 – belts
1 – insect net
also, 30% of a unit's personnel was issued the tropical helmet

NOTE:
The mixing of blue-grey and tropical uniform parts was forbidden.

[10] Luftwaffen-Verordnungsblatt, September 1, 1941. The order is numbered 1090 and dated August 16, 1941.

[11] Luftwaffen-Verordnungsblatt, February 21, 1944. The order is numbered 232 and dated February 11, 1944.

The Great-Coat

The basic design of the blue-grey Great-Coat was the same for all ranks and worn with collar patches and shoulder straps.[12] It could be worn with open or closed collar. Great-Coats for general ranks varied to an extent with their lapels being in Waffenfarbe (white for generals, dark green for officials of general's rank).

Officers could also wear a coat in the same cut and color as the Great-Coat, but in rubberized material for rain. A blue-grey leather coat was also authorized. The rain-coat and leather coat were worn with detachable shoulder straps, but without collar patches. The Great-Coat had two rows of six buttons which were in aluminum as for the Service Tunic (gold for generals). A sword or dagger hanger was suspended from beneath the left pocket flap.

General's and officer's Great-Coats

(Left) General Christiansen wearing a leather Great-Coat. Note car pennant for Chef einer Luftflotte in background.

[12] Luftwaffen-Verordnungsblatt, May 29, 1942. This order numbered 1423, stated that the wearing of collar patches was to be discontinued by NCOs and men from October 1, 1942 and by owners of private purchase Great-Coats (officer ranks) from April 1, 1943. All Luftwaffe units followed this ruling, with the exception of the Wachregiment in Berlin and the Führer-Flakabteilung of the Hermann Göring Division.

The Luftwaffe Cape

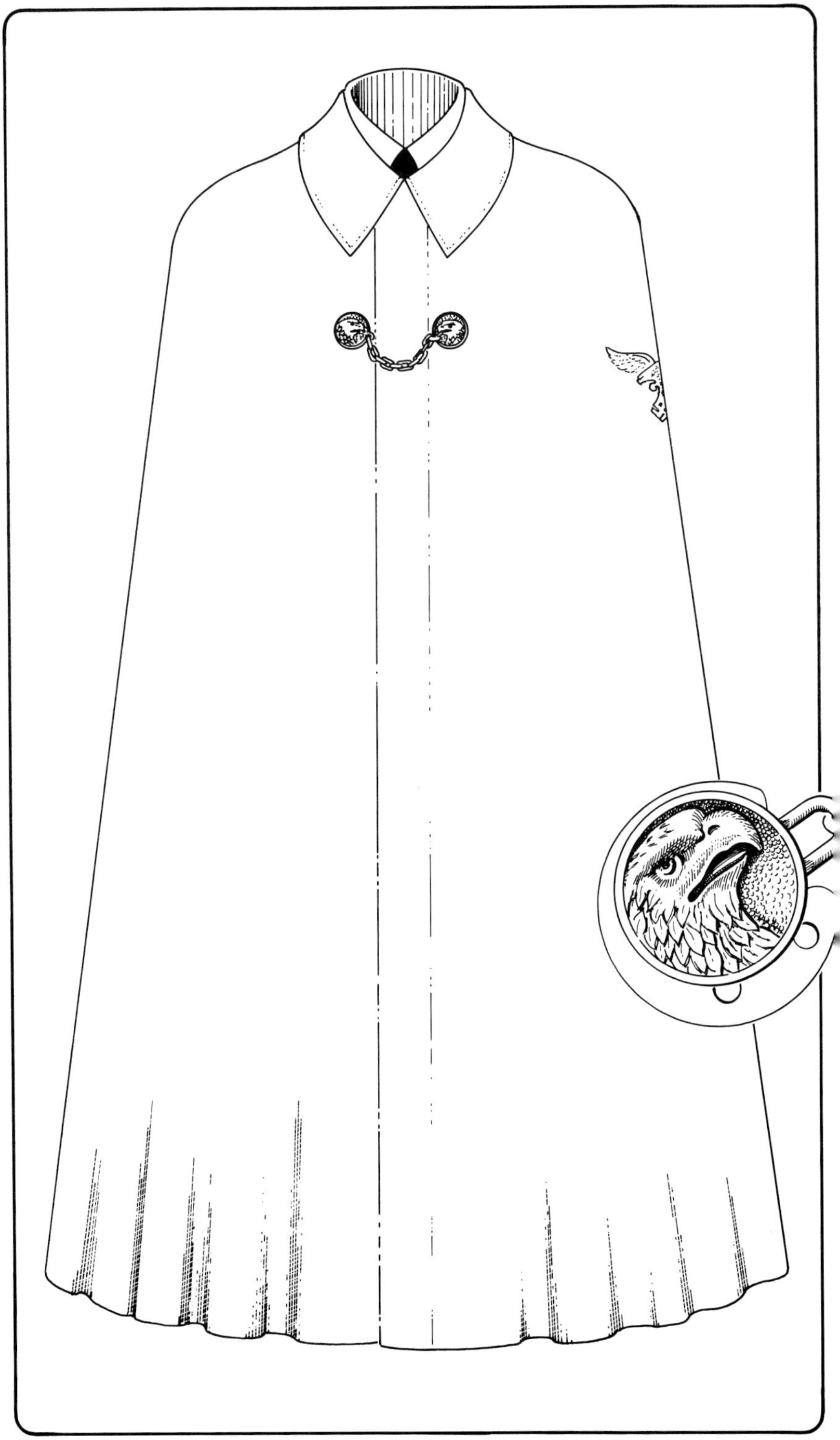

The Luftwaffe cape was available in a blue-grey cloth or rubberized material. Its wear was permitted by officers, plus officials, technicians and navigational experts of officer rank.[13] When worn, the bottom of the cape was to have been one hand's width below the knee.

A clasp in the form of two facing eagles' heads attached by a chain, was positioned on the breast (see illustration below). It was in silver-colored metal (gold for generals). A large (17.5mm wide) Luftwaffe eagle was positioned on the left upper arm of the cape in hand-embroidered aluminum wire for officers and in gold wire for generals. No eagle was worn on the rubberized version.

Generals also had their collar lining in white as well as a white piping around the collar (officials in general rank – dark green, technicians in general rank – pink).

Providing a definite dress had not been prescribed for the day, the cape could be worn with all uniforms. It could also be worn over Great-Coats.[14]

NOTE:
After the outbreak of war in 1939, the cape was worn less and less and finally disappeared from use.

Note Göring's hand-embroidered Reichsmarschall eagle on the upper left arm.

Luftwaffe cape clasp for officers (silver) and generals (gold)

[13] Also those officials allowed to wear the portepee plus aluminum chin straps on their peaked cap.

[14] Anzugordnung für die Luftwaffe, November 27, 1935, revised on April 1, 1937, pp. 35-37.

Waterproof Coat for Motorized Units

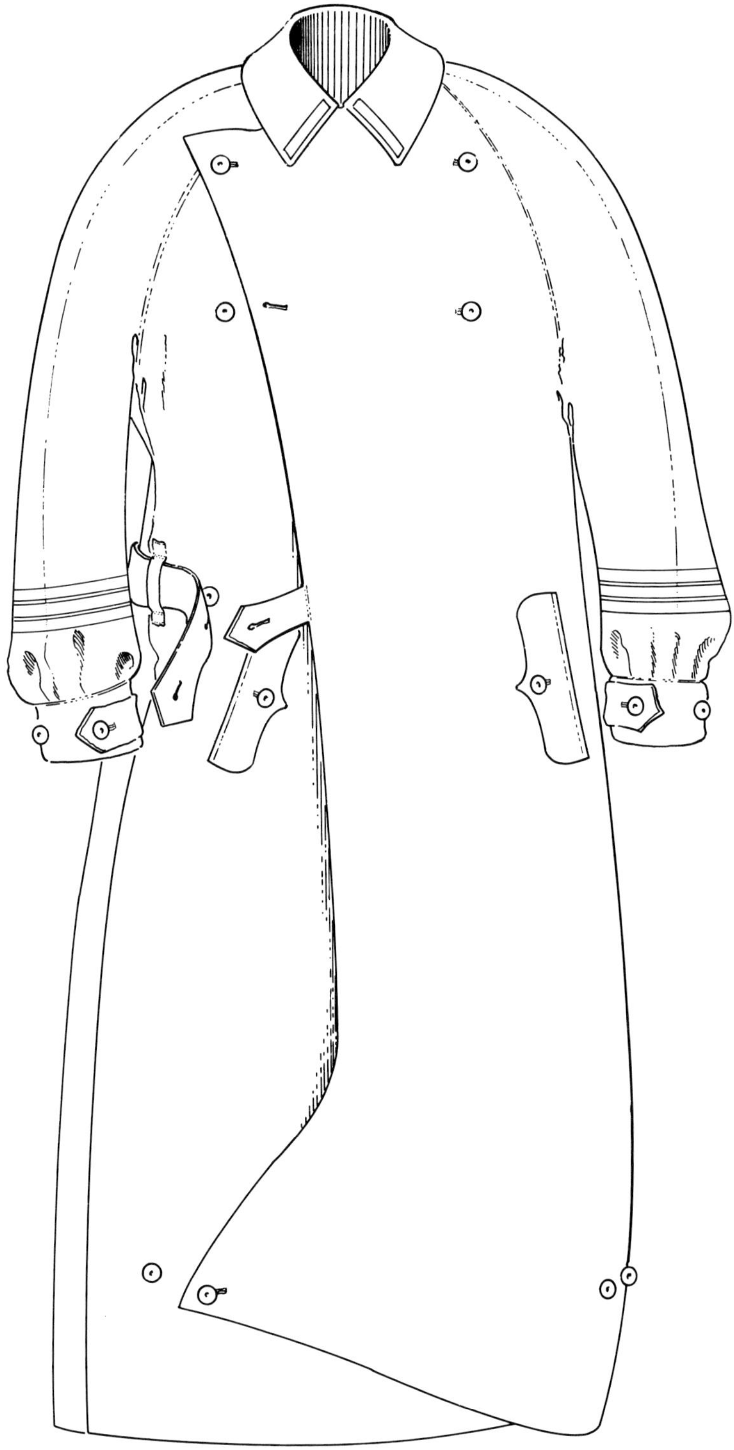

The motoring coat was of a blue-grey rubberized material and identical in design to that of the Army's. It did not, however, have accomodations for shoulder straps as did the Army's. Rank insignia for Luftwaffe NCOs was a grey braid or "Tresse" on the front of the collar as well as one, two or three sleeve stripes for Feldwebel, Oberfeldwebel, Hauptfeldwebel and Stabsfeldwebel (or Wachtmeister, etc.) respectively (same as the sleeve ranks for the fatigue uniform). Rank designations for officers were as on the Flying Suits. Chevroned grades wore no insignia on this coat.

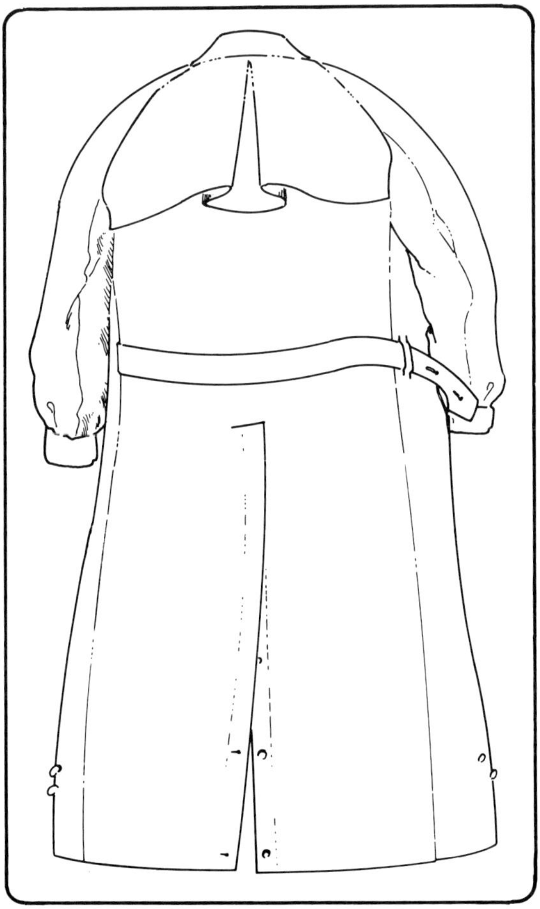

Obverse

Stabfeldwebel/
Stabswachtmeister

Hauptfeldwebel/
Hauptwachtmeister

Oberfeldwebel/
Oberwachtmeister/
Oberfähnrich

Feldwebel/Wachtmeister

Motoring Coat for a Luftwaffe Hauptfeldwebel

Luftwaffe Flying Suits

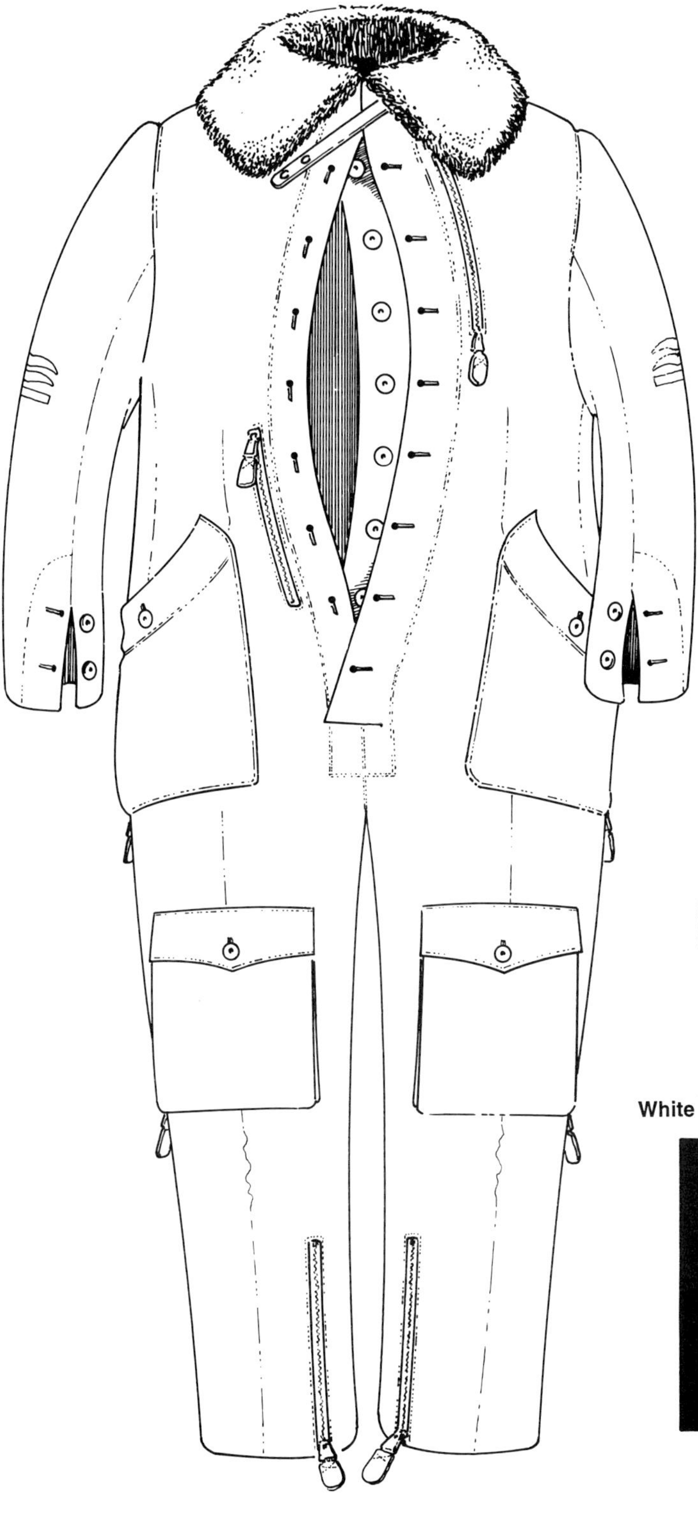

1

White batons

5

Three major variations of the Flying Suit were in service: a light-weight summer suit for all terrain, a suit for winter over land (illustrated) and a suit for winter over sea.

The light-weight summer Flying Suit was in a tan material and worn directly over the Flight Blouse and pants. Officers and NCOs utilized the ranking system shown below and on the next two pages. The double wing and bar system was white (yellow for generals) on a tan base for this suit and worn on both arms. Enlisted men wore no rank designation on their Flying Suit. An unlined flying helmet was worn with this suit.

The winter Flying Suit for flights over land was more heavily constructed with a cotton or fleece lining. It was in either tan, dark-blue or dark-grey material with its rank insignia in white (yellow for generals) on its appropriate base color.

The winter Flying Suit for flights over sea was again heavily constructed, but of brown or black leather, and cotton or fleece lined. Its rank insignia was white (yellow for generals) on a thin brown or black leather base. Lined flying helmets were worn with the two winter suits.

For operations in extreme cold, an electrically heated suit, gloves and foot coverings were worn over the Flight Blouse and pants, and under the Flight Suit.

2

3

4

6

7

8

A bomber crew wearing the summer Flying Suit (Note position of rank insignia).

9

10

11

15

16

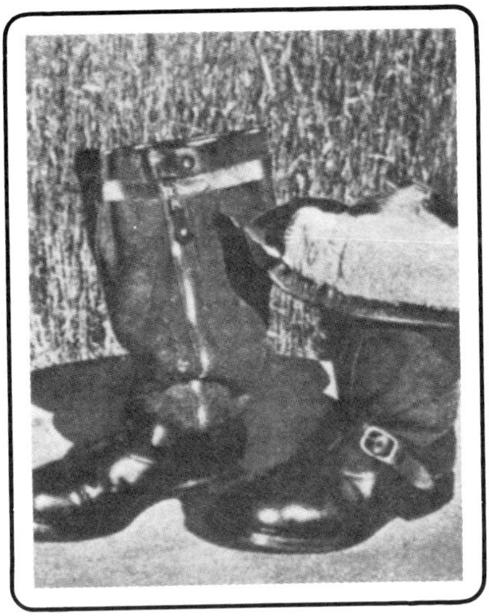

Flying boots were fleece-lined and of black leather or a combination cloth and leather (above).

Flight boots were frequently worn without Flight Suit by flying personnel.

12

13

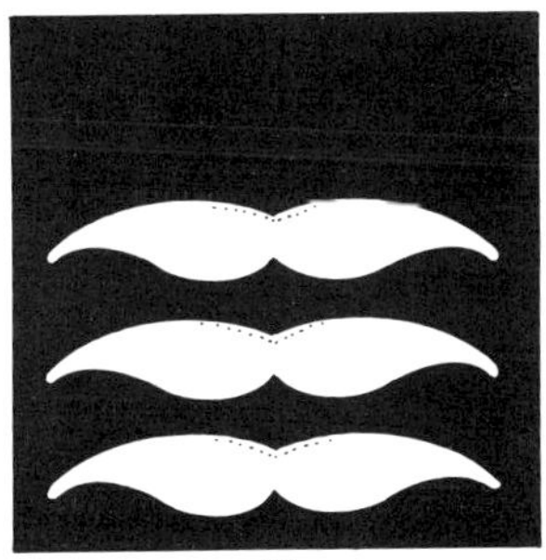

14

1. Generalfeldmarschall
2. Generaloberst
3. General der Flieger
4. Generalleutnant
5. Generalmajor
6. Oberst
7. Oberstleutnant
8. Major
9. Hauptmann
10. Oberleutnant
11. Leutnant
12. Stabsfeldwebel
13. Oberfeldwebel/Oberfähnrich
14. Feldwebel
15. Unterfeldwebel/Fähnrich
16. Unteroffizier

Life Jackets

Life jacket for crews of bombers, transports and flying boats.

Life jacket for fighter pilots and crews of two-seater aircraft.

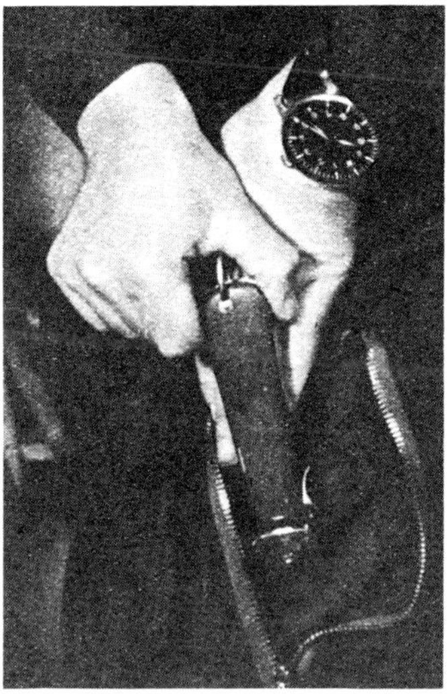

The utility knife above, was issued as part of the Luftwaffe air crewmen's survival equipment and to all parachute troops.

Cartridges for the flare pistol were strapped on, below the knee.

Working Uniform (Fatigue)

These uniforms, which were available in one or two pieces, were of a course light grey, dark blue-grey or black twill, usually unlined and without any branch of service designation. The two-piece Fatigue Uniform was issued in light grey for drill and training purposes, whereas the Work Uniform (one or two piece) was normally issued in the darker colors to better conceal dirt and grease.[15]

Enlisted men's ranks were designated by the wearing of rank chevrons which were on a blue-grey backing, as on the Service Tunic or Flying Blouse. NCOs wore the ranking system illustrated below. The NCO collar "Tresse" for both the Fatigue and Work Uniform was in Luftwaffe blue-grey. Officers wore both the Fatigue and Work Uniform but with the wing and bar insignia (as on the Flying Suit) on both arms with a black underlay.

An order dated June 4, 1940, stated that only course grey fatigues would be issued for the duration of the war, and work suits, lined and unlined, plus work forage caps (only upon request) would be issued in dark blue-grey or black.[16]

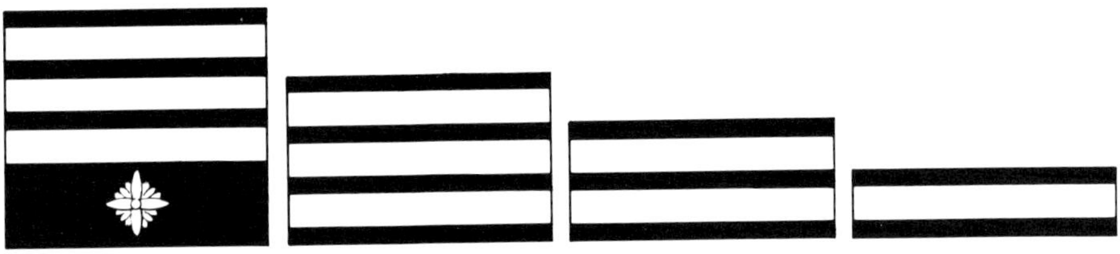

Stabsfeldwebel/ Stabswachtmeister | **Hauptfeldwebel/ Hauptwachtmeister** | **Oberfeldwebel/ Oberwachtmeister/ Oberfähnrich** | **Feldwebel/ Wachtmeister**

NOTES:

1. An order dated late September 1937, stated that Fatigue and Work Uniforms would now be issued with only a right breast pocket... initially there were pockets on both breasts. A number of these uniforms did, however, exist with no breast pockets.
2. During the winter, the Fatigue and Work Uniform could be worn over the Flight Blouse and pants.

(Above) Feldwebel in fatigue coveralls.

Young recruits receiving instuctions on the "goose-step."

[15] This dark-grey or black twill coverall with a fly front was furnished to individuals whose job required them to become extremely dirty, such as flak crews and aircraft mechanics.

[16] Luftwaffen-Verordnungsblatt, June 17, 1940. This order is numbered 708.

Flying Jackets

Pilots of fighter aircraft required as much mobility as possible in their confining cockpits. Therefore in mid-1940, they were permitted to wear a light-weight jacket or windbreaker in place of a Flying Suit.[17] The jackets took many forms and designs, as they were largely private purchase items. Their colors varied from shades of grey to black. A few variations are illustrated below.

Oberleutnant Erich Hartmann, holder of the Knight's Cross with Oakleaves, Swords and Diamonds.

Major Werner Mölders

[17] Luftwaffen-Verordnungsblatt, July 15, 1940. This order is numbered 876 and dated July 4, 1940.

Hauptmann Zemsky, a Stuka pilot, with his wireless operator after their 400th combat mission.

The Sports Kit

The Sports Kit consisted of a white pull-over sportshirt, dark blue cotton trunks and running shoes. A 24 x 11.5 cm Luftwaffe eagle in black on white was positioned 7 cm below the neck opening, about mid-breast. The badge of rank for the sport shirt was a 1 cm wide black twill stripe around the neck for NCOs, and a 1 cm wide black twill stripe plus another separate stripe, 0.5 cm wide, for officers. The Staffel of individual enlisted men was indicated by a 12 cm wide band running around the breast directly below the Luftwaffe eagle (red, yellow, blue or green).[18]

For those Luftwaffe personnel who were members of the Luftwaffe Sports Association, the Wehrmacht sport emblem on a colored base was initially worn directly below the Luftwaffe eagle on the colored Staffel band.[19] It was later worn on the lower left breast below the band.

[18] Luftwaffen-Verordnungsblatt, July 15, 1935. This order was numbered 366 and dated July 4, 1935.

[19] Luftwaffen-Verordnungsblatt, March 2, 1936. This order was numbered 248 and dated February 19, 1936.

Sports (Training) Suit

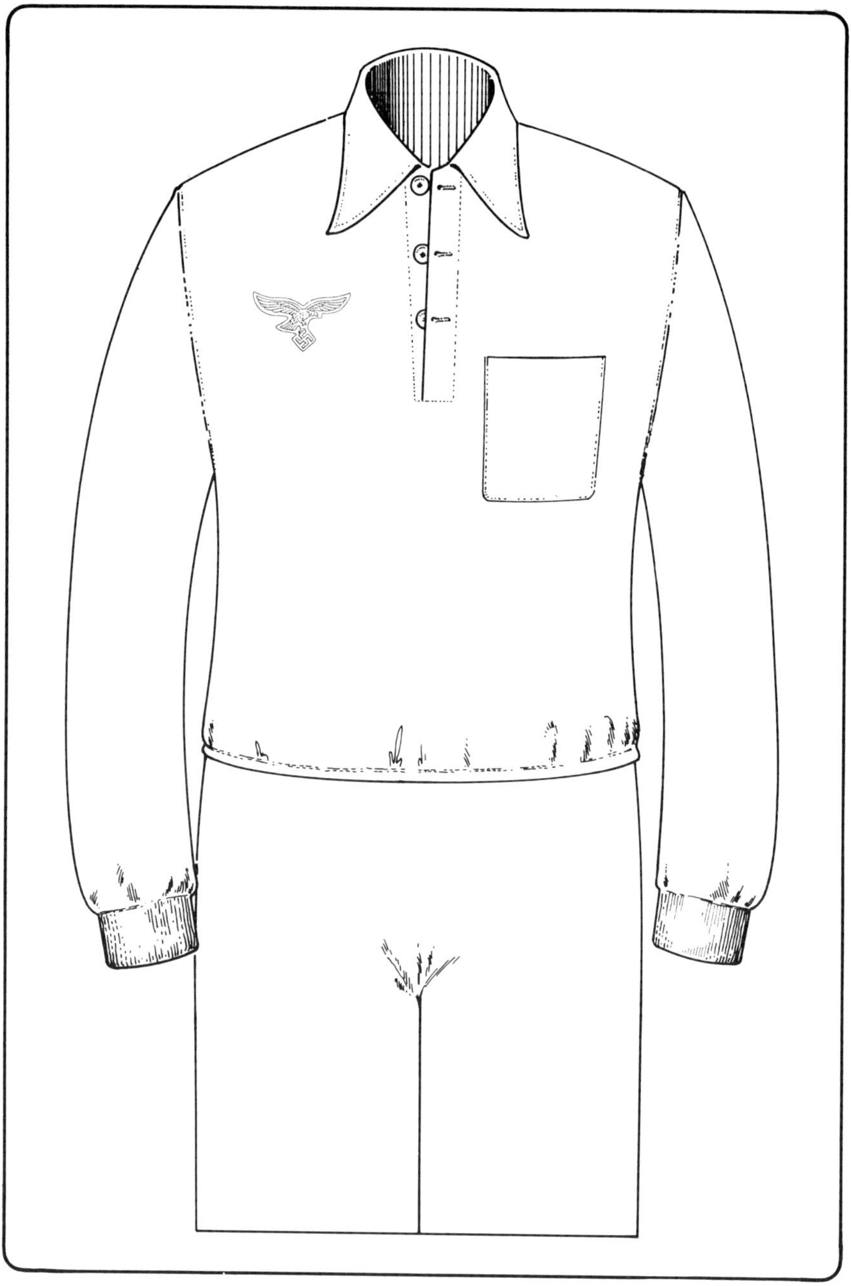

The two-piece Sports Suit was in a dark navy blue material and worn usually while participating in football or handball. The collar was cut allowing the pull-over blouse to be worn open or closed. On the left breast, was a 16 x 22 cm pocket and on the right breast a matte-grey embroidered Luftwaffe eagle on a dark blue backing.[20] Rank insignia for this suit was worn on both arms (between the shoulder and elbow), and was as that worn on the Flying Suit and Work Uniform by NCOs and officers (white insignia on a black backing). Wehrmachtbeamte personnel wore this suit with the arm insignia as above but with triangular pips instead of the double wings.

With the Sports or Training Suit, a blue-grey ski cap (styled after the Army's "Bergmütze") was worn with the Luftwaffe eagle in matte-grey embroidery.

Swimming Trunks for Luftwaffe Personnel

The trunks were in black for bathing and swimming school and in a light green for experienced swimmers (Freischwimmer) and life guards. Life guard teachers (those who have passed their examination) have a 1.5 cm wide, white stripe around the top of the trunks.

[20] Luftwaffen-Verordnungsblatt, July 27, 1936. This order was numbered 89 and dated July 18, 1936.

Headgear

The Peaked Cap

The Service or Uniform Cap for enlisted men was a peaked cap with a blue-grey worsted cloth/30% wool cover while that of the officers was of a finer material. The cap band was 4.3 cm wide, of a black mohair for all ranks and had a 2 mm wide piping on top and bottom thusly:

(a) for NCOs, NCOs entitled to wear officer's cap cords and other ranks – appropriate Waffenfarbe

(b) for all officers – aluminum

(c) for all general ranks – gold

The above also refers to a 2 mm wide piping which was around the crown of the cap. The Service Cap was worn with the following orders of dress:

(a) with Reporting and Walking-Out Dress for NCOs and other ranks
(b) with Service Dress and Undress uniform for officers and NCOs
(c) with Parade Dress, by officers when not actually on duty
(d) with Walking-Out Dress and Dress uniforms by officers

An Obergefreiter wearing the Service Cap for enlisted men with metal insignia.

Oberst Mölders wearing the officer's Service Cap with hand-embroidered insignia.

Hand-embroidered insignia for the officer's Service Cap.

Cap cords for the Service Cap were in a gold color and fastened with gold buttons for general ranks, in aluminum with aluminum colored buttons for officers, and in black leather with black buttons for NCOs and men. The national emblem, which was positioned on the front of the cap cover, was in a hand-embroidered gold wire for generals, a hand-embroidered aluminum wire for officers, and in a stamped aluminum colored metal for NCOs and men. The cockade, which was surrounded by oak leaves and flanked by wings, was also in hand-embroidered gold wire for general ranks, in hand-embroidered aluminum wire for officers, and in a stamped aluminum colored metal for NCOs and men. It was postioned on the black mohair cap band directly below the national emblem.

Oberleutnant Beisswenger wearing his Service Cap without the rigid cap spring.

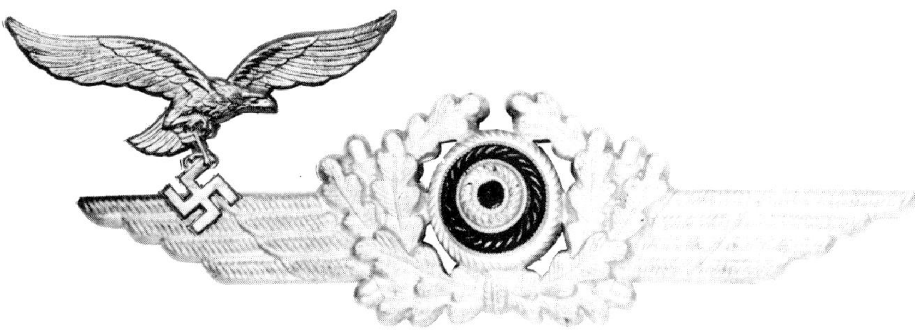

Metal insignia for the enlisted man's Service Cap.

Old Style Officers Service Cap

Officers could also wear an alternative field service cap known as the Old Style Officers Field Service Cap. It was similar in appearance to the standard Luftwaffe Service Cap but without officers cap cords, without a rigid cap spring and with a soft leather visor. The insignia was usually a less elaborate embroidered version.[1]

Summer Service Cap

The Summer Service Cap was composed of a firm framework which did not allow it to be folded or crushed. The summer version was as the standard peaked cap with the exception that its cover was in white and removeable. The piping around the crown of the cap was white for officers, and in appropriate colors for NCOs and men.

Officer in the background is wearing the white Summer Service Cap authorized between the dates of April 1 and September 30. Generaloberst Rommel is in the foreground.

[1] Anzugordnung für die Luftwaffe, p. 14. Aluminum hand-embroidered insignia (national emblem and winged cockade) was officially designated for officers up to Oberst, and included Oberfähnriche, Unterärzte-, Obermusikinspizient down to Musikmeister, Oberfeuerwerker (after examination for promotion to officer rank), all officials of officer rank and Beamtenanwärter (candidate for official in administrative service after examination) in the rank of Oberfeldwebel, Obermeister and Meister in the Reichsluftaufsicht and the ranks of Hauptbrandmeister and Brandmeister as Feuerschutzbeamte der Luftwaffe.

Tropical Service Cap

This peaked cap (introduced for wear on 13.4.42), which could be worn in place of the tropical Forage Cap, was for officers, NCOs and men who were issued the tropical uniform. It was in tan with a cloth visor and embroidered insignia on a tan base. A removeable neck protector was also issued with this cap.

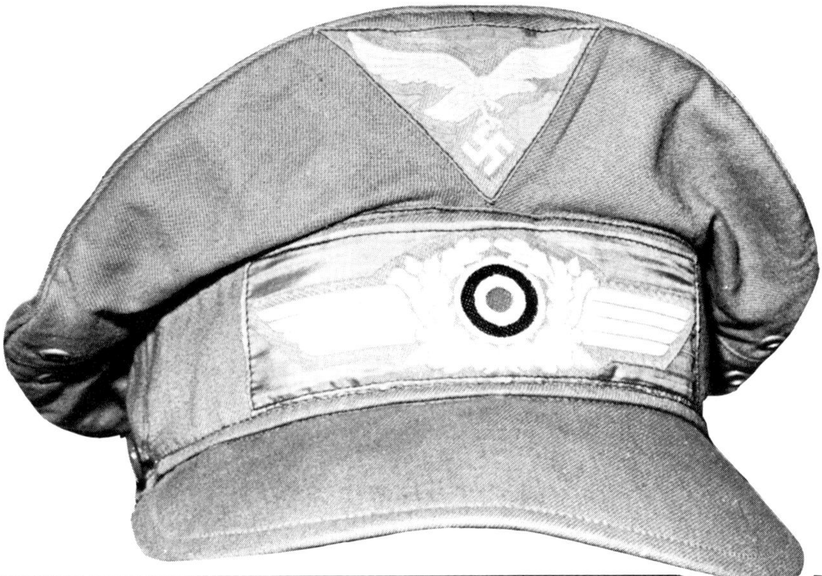

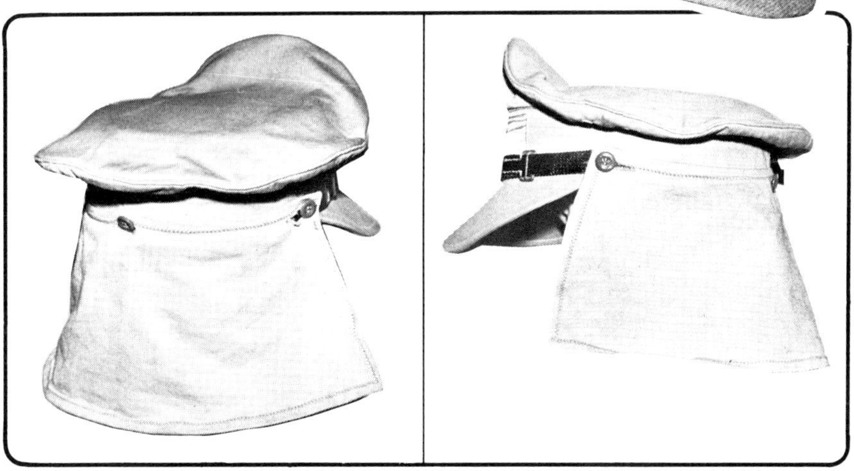

NOTES:

1. The Luftwaffe eagle and Reich cockade were introduced for wear on August 27, 1935, thus replacing the Luftsportverband insignia which had been worn up to this time.

2. Although forbidden, many Luftwaffe personnel of all ranks removed the rigid cap spring from inside the cover, resulting in the so-called "Luftwaffe crush."

The Forage Cap

Major Lützow

The Luftwaffe Forage Cap was worn for occasions not covered by the steel helmet or peaked cap. The blue-grey cap was of the same cut for all ranks and service branches within the Luftwaffe (see illustrations). The Luftwaffe national emblem in matte-grey embroidery for NCOs and men, aluminum embroidery for officers and gold colored embroidery for generals, was positioned in the middle of the vertical seam in front, with the swastika coming to the upper edge of the turn-up. Directly below the Luftwaffe eagle, in the middle of the turn-up, was the national cockade in black/white/red.

Officers, engineering officials, nautical officials, officials of officer rank plus those NCOs allowed officer's cap cords wore a Forage Cap of a finer material and with a 3 mm wide aluminum cord around the upper edge of the turn-up. For general ranks (also officials), a 3 mm wide gold colored cord was around the upper edge of the turn-up.

Regulations stated that this cap was to be tilted towards the right ear so that the bottom edge of the cap was one finger's width above the ear. The cockade was to be aligned with the center line of the face.[2]

Hand-embroidered officer's cap eagle

Machine-embroidered enlisted man's cap eagle

Enlisted man's Forage Cap

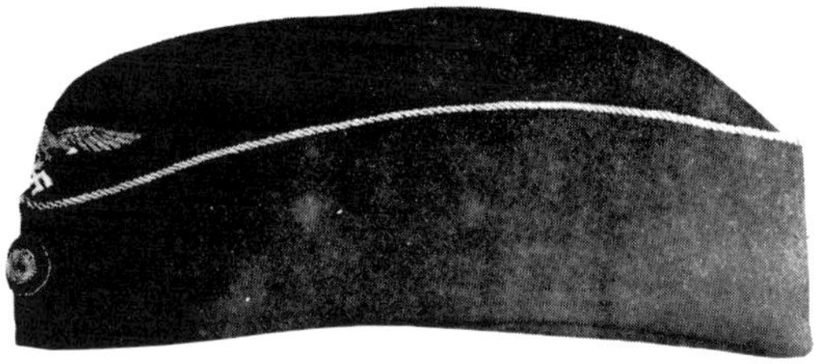

Officer's Forage Cap

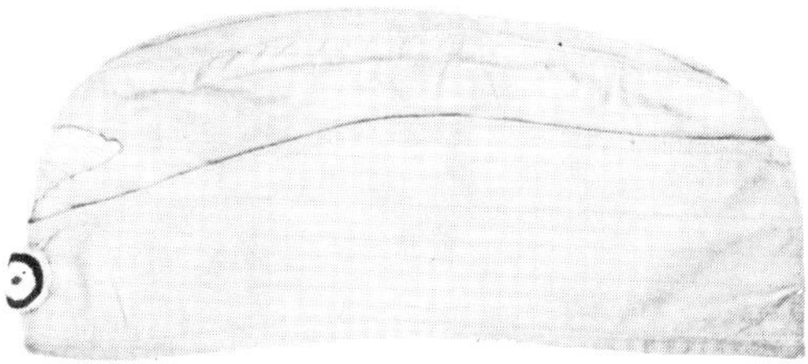

Tropical enlisted man's Forage Cap

Black Forage Cap (work uniform)

NOTES:

1. The Reich cockade (black/white/red) was positioned on the Forage Cap below the Luftwaffe eagle for all ranks, except for the black forage cap in black twill (work uniform) which had only the Luftwaffe eagle on a black base.
2. The Forage Cap also existed in a tan for wear with the tropical uniform and had the embroidered Luftwaffe eagle and cockade on a tan base.

[2] Anzgordnung für die Luftwaffe, p. 12.

The M1943 Cap

A visored cap, whose design was based on the Army's mountain cap, was introduced for wear by all Luftwaffe troops in 1943. The cap was plain and unpiped for NCOs and men. Officers and officials of officer rank were authorized to wear the mountain-style cap with a 3 mm aluminum piping around the top edge of the cap (gold for general ranks). Variations exist with the scalloped portion of the turn-up also being piped.

The insignia consisted of a matte-grey embroidered Luftwaffe eagle for NCOs and men, aluminum emboidery for officers and gold embroidery for generals. An embroidered national cockade (black/white/red) was positioned directly below the Luftwaffe eagle. Initially, the cap was issued with two blue-grey, plastic or metal buttons on the M1943 turn-up, which were later replaced with one 12 mm blue-grey, pebbled, metal button (aluminum for officers and officials in officer's rank plus gold for generals and officials of generals' rank).

NOTES:

1. Although mountain units of the Army and Waffen-SS were authorized to wear the distinctive Edelweiss emblem on their M1943 caps, members of Luftwaffe mountain units were forbidden to wear such special insignia on their caps or anywhere else on their uniform.[3]
2. The M1943 Cap was introduced to eventually replace the Forage Cap.
3. The M1943 Cap was also issued in tan for wear with the tropical uniform.

[3] Luftwaffen-Verordnungsblatt, September 4, 1944. The order was numbered 1187 and dated August 23, 1944.

The Luftwaffe Steel Helmet

Luftwaffe helmets, which were issued from 1935-1945, were the old transition model (W.W.I style with M31 liner plus decal), the Model 1935 (with rolled edge) and the Model 1942 (with outward crimped edge).

The helmet was painted inside and out with a rust-resistant blue-grey paint. A shield in the national colors (black/white/red) was positioned 3 mm below the ventilation hole on the right side of the helmet and the Luftwaffe national emblem in silver-grey on the left side.[4] For generals, as well as technicians and officials of general rank, the Luftwaffe eagle was in gilt.

Luftwaffe personnel manning a light machine-gun in defense of a wireless station (Poland).

National emblem decal (Pre-war transfer)

National emblem decal (Transfer)

[4] When in combat zones, the national colors were ordered removed in early 1941.

The Helmet Camouflage Net

As the war progressed and more and more Luftwaffe units became involved in ground combat, a camouflage helmet netting was made available for wear. It was available in early September 1942 and was attached to the helmet as follows:

> A front hook was positioned between the outer ring of the helmet liner and the helmet itself. A rear hook was attached between the inner and outer ring of the helmet liner. The netting was attached to the front hook at point "a" and affixed to the helmet as illustrated.

Camouflage material which could be suitable for use with the netting were dry leaves, dry fern, moss, short grass blades, leaves of beets, heather, wheat, reeds, pieces of bark and pieces of cloth. In order to make the camouflage more effective, it was recommended that the camouflage material be pulled under the netting (which had been pulled tight by means of a tension cord). Small twigs could be pulled under the tension cord in a sideway fashion. This was one method of breaking up the smooth silhouette of the helmet and blending it into the surrounding terrain for improved concealment.[5]

Luftwaffe personnel riding the fenders of an official auto, and searching the skies for Allied fighter-bombers (Normandy).

[5] Luftwaffen-Verordnungsblatt, September 14, 1942. The order is numbered 2326 and dated September 3, 1942.

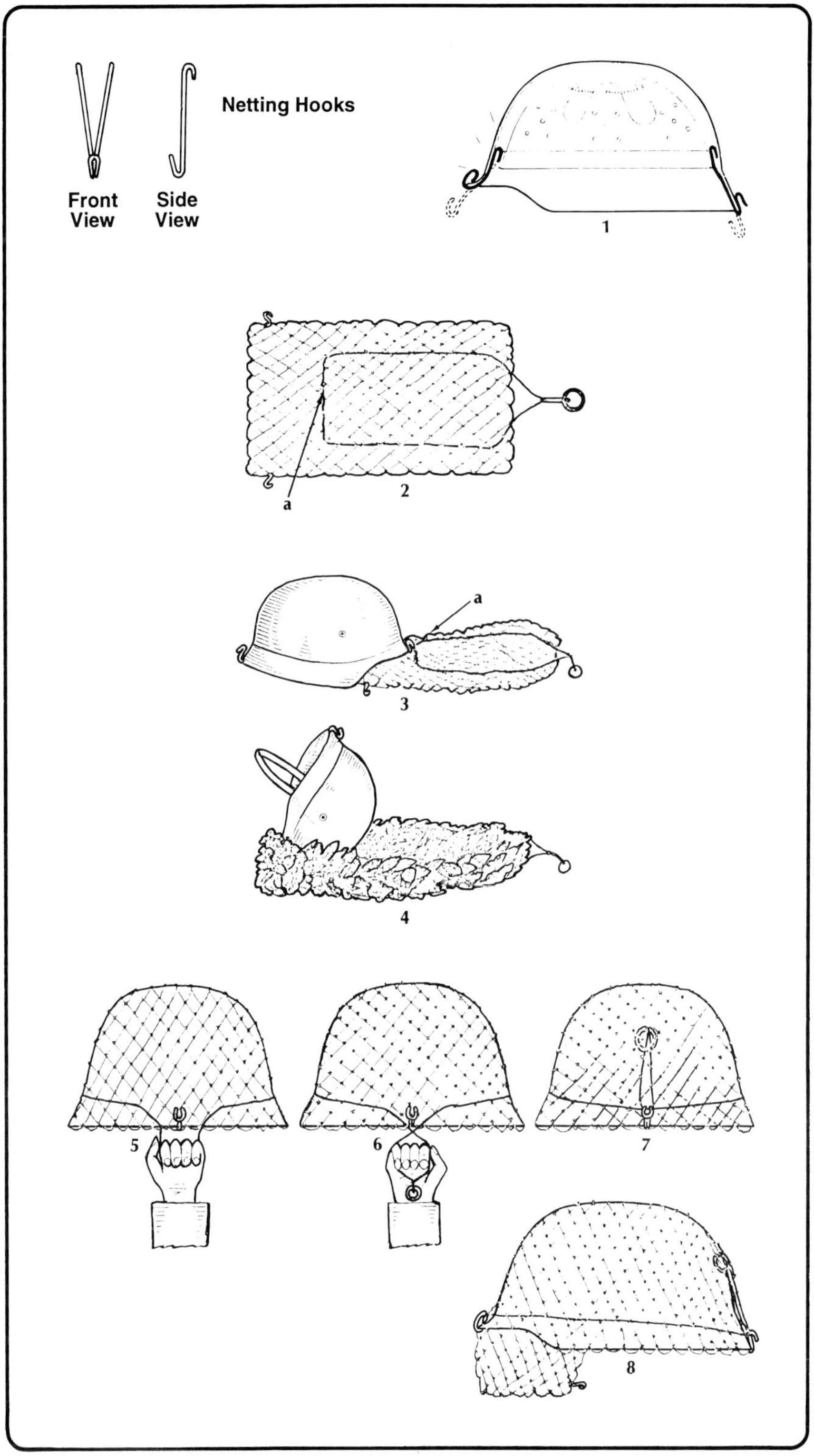
Netting Hooks
Front View
Side View
1
2
a
3
a
4
5
6
7
8

The Parachutist Helmet

This specially designed helmet was of a compact design with few protruding parts. It had a unique rubber padded lining and an unconventional leather chin harness.[6] The early issue helmets were of a blue-grey color with the national colors (black/white/red) on the right side and a silver-grey Luftwaffe eagle on the left. The national colors were, however, ordered removed early in the war. As the war progressed, the color of the helmet was changed to a dark-green. For parachute units in Russia, home-made white covers were utilized or the helmets were painted white. When fighting in North Africa or in the Mediterranean area, the helmet was painted a tan color. Mud coverings were also applied for camouflage purposes.

Issue helmet covers were utilized to a great extent and were initially of a blue-grey material. There existed also two variations of a camouflage pattern cover. Steel "chicken wire" covers and large holed netting were additionally used in the parachutist's camouflage techniques.[7]

Note the helmet in forground which was painted white. The commander of the 1. Fallschirmjäger Division, Generalleutnant Heidrich is on the right.

[6] When the parachutist was removed from jump status and utilized strictly as an infantryman, the special chin harness could be replaced by a D-ring and a standard helmet buckle (if time and the situation allowed).

[7] Helmets were also painted in camouflage patterns.

(Above) Note home-made camouflage cover in foreground and large holed netting in background. (Below) 1st issue camouflage cover.

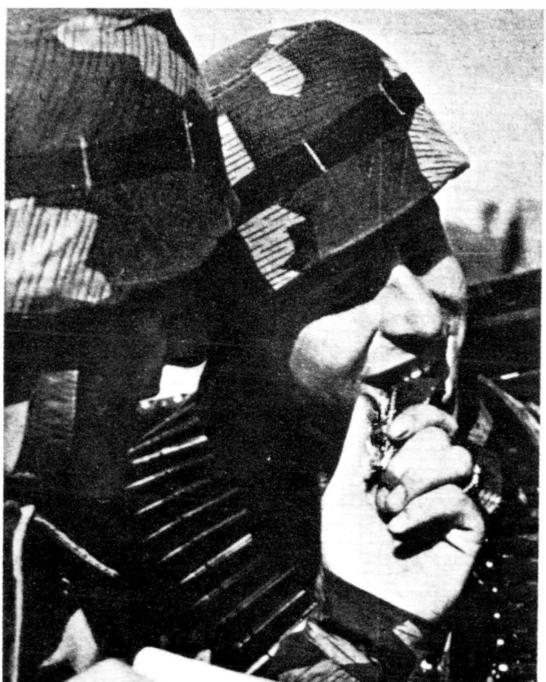

National emblem for left side of helmet

National colors for right side of helmet

(Above) Blue-grey helmet cover

The Flying Helmet

There were three major variations of the Flight Helmet: a summer Flying Helmet, an unlined Flying Helmet and a fleece-lined Flying Helmet.

Unlined Flying Helmet

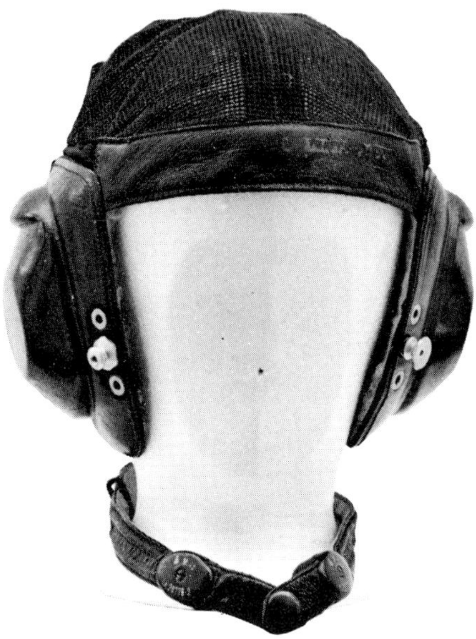

Summer Flying Helmet

Fleece-lined Flying Helmet

The SSK 90 Steel Flight Helmet

The SSK Steel Flight Helmet was issued to some flight crews on an experimental basis in 1941. It was, however, considered unacceptable for service and returned to the Luftwaffe Clothing Office in Berlin. It was constructed of brown leather-covered steel, with a rubber padded comb and cut-outs on each side of the helmet for earphones.

The SSK 90 Steel Flight Helmet

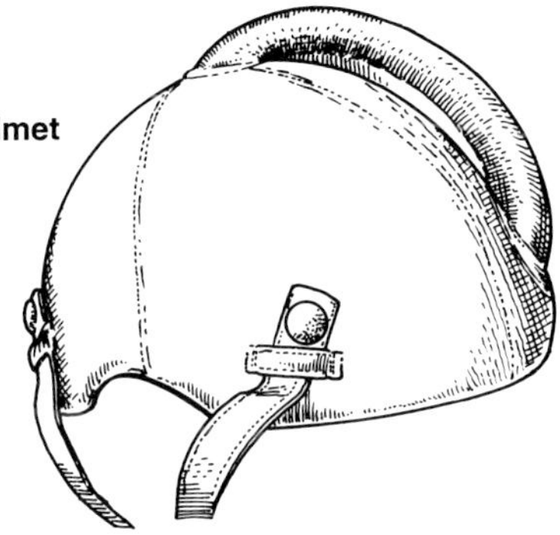

The Tropical Helmet

The Luftwaffe Tropical Helmet was in tan (some in blue-grey) and constructed identical to that of the Army's. The insignia was an embossed, metal national shield (black/white/red) on the right side and a metal Luftwaffe eagle on the left.

Winter Fur Caps

Various patterns of the Winter Fur Cap, issued to Luftwaffe personnel, were encountered. Some had the metal Luftwaffe eagle attached to the front portion of the cap while others employed the embroidered eagle. A few of the variations are illustrated below.

Generalmajor Adolf Galland

Cuff Titles and Armbands

Cuff Titles

Immediately upon breaking the binding chains of Versailles, the Luftwaffe embarked on a policy of rebirth of World War I traditions which was to inject pride and enthusiasm into the new air arm. Heroes of the last war and also of the Nazi Party were commemorated with the renaming of squadrons and issuance of commemorative cuff titles (i.e. "Geschwader Immelmann"). These cuff titles were worn 1 cm above the right cuff of the service tunic, flying blouse and summer tunic by officers, and on the service tunic and flying blouse for NCOs and men.

The cuff titles were of a dark blue base with the commemorative inscription in aluminum embroidery for officers and matte-grey embroidery for NCOs and men.

Jagdgeschwader Richthofen

By order of the Führer and Supreme Commander of the Wehrmacht, on March 14, 1935, a new 3.3 cm wide commemorative cuff title "Jagdgeschwader Richthofen" was to be worn by all officers, NCOs and men of "Fliegergruppe Döberitz",[1] which was later Jagdgeschwader 132, the JG 131 and finally JG 2.

Major Helmuth Wick, commander of Jagdgeschwader Richthofen, and Generalfeldmarschall Sperrle.

[1] Luftwaffen-Verordnungsblatt, April 8, 1935. The order is number 91 and dated March 25, 1935.

Sturzkampfgeschwader Immelmann

By order of the Führer and Supreme Commander of the Wehrmacht, on March 14, 1935, a new commemorative cuff title "Geschwader Immelmann" was to be worn by all officers, NCOs and men of "Fl. Gruppe Schwerin",[2] which later became Sturzkampfgeschwader 162 and finally Sturzkampfgeschwader 2.

Kampfgeschwader Boelcke

By order of the Führer and Supreme Commander of the Wehrmacht, a new commemorative cuff title "Geschwader Boelcke" was to be worn by all officers, NCOs and men of "Fl. Gruppe Fassberg",[3] which later became Kampfgeschwader 154, then KG 157 and finally KG 27.

Hauptmann Kilian Koll

NOTE:

Those belonging to Geschwader Richthofen, Immelmann and Boelcke, and who had served in the flying unit of the same name in W.W.I, were permitted to wear a 3 mm wide aluminum tresse on the top and bottom of their cuff title.[4] This practice was, however, discontinued at a later date.

[2] Luftwaffen-Verordnungsblatt, May 6, 1935. The order is numbered 164 and dated April 23, 1935.

[3] Ibid.

[4] Ibid.

Zerstörergeschwader Horst Wessel

By order of the Führer and Supreme Commander of the Wehrmacht, on March 24, 1936, a new commemorative cuff title "Geschwader Horst Wessel" was to be worn by all officers, NCOs and men of "Fliegergeschwader Dortmund",[5] which became Zerstörergeschwader 142 and finally ZG 26.

Kampfgeschwader Hindenburg

By order of the Führer and Supreme Commander of the Wehrmacht, on April 20, 1936, a new commemorative cuff title "Geschwader Hindenburg" was to be worn by all officers, NCOs and men of "Fliegergeschwader Greifswald",[6] which later became Kampfgeschwader 152 and finally KG 1.

Kampfgeschwader General Wever

By order of the Führer and Supreme Commander of the Wehrmacht, a new commemorative cuff title "Geschwader Wever" was to be worn by all officers, NCOs and men of "Kampfgeschwader Gotha",[7] which later became Kampfgeschwader 253 and finally KG 4.

[5] Luftwaffen-Ordnungsblatt, April 15, 1936. The order is numbered 457 and dated May 6, 1936.

[6] Luftwaffen-Ordnungsblatt, May 18, 1936. The order is numbered 586 and dated May 6, 1936.

[7] Luftwaffen-Verordnungsblatt, June 8, 1936. The order is numbered 686 and dated June 4, 1936.

Jagdgeschwader Frhr. von Richthofen Nr. 1 1917/18 and Jagdstaffel Boelcke Nr. 2 1916/18

The qualifications for the wearing of the above commemorative cuff titles were stated by Generaloberst Göring. Those who had been an honored member of "Jagdeschwader Richthofen Nr. 1" (WWI title) and "Jagdstaffel Boelcke Nr. 2" (WWI title) were entitled to wear the cuff title but with the following stipulations:

(a) Proof must be given of an individuals membership in one of the two units, either through visual recognition or by producing the proper "Kriegsrangliste" (listing of personnel in various units),

(b) and must have served in the unit for a minimum of one year (after proving affiliation), or

(c) while a member of the unit in question, to have been wounded.[8]

Generalfeldmarschall von Richthofen (wearing the "Jagdgeschwader Richthofen Nr. 1" cuff title) and Generalfeldmarschall von Manstein.

[8] Luftwaffen-Verordnungsblatt, June 21, 1937. The order is numbered 751 and dated June 11, 1937.

Jagdgeschwader Schlageter

By order of the Führer and Supreme Commander of the Wehrmacht, a new cuff title "Jagdgeschwader Schlageter" was worn by all officers, NCOs and men of "Jagdgeschwader 132," which later became JG 26.[9]

Oberst Galland

Legion Condor (traditional)

By order of the Führer and Supreme Commander of the Wehrmacht, a new traditional cuff title "Legion Condor" was to be worn by all officers, NCOs and men of:[10]

(a) "Kampfgeschwader 53"
(b) "Flakregiment 9"
(c) "Luftnachrichten-Regiment 3"

[9] Luftwaffen-Verordnungsblatt, January 16, 1939. The order is numbered 48 and dated December 8, 1938.

[10] Luftwaffen-Verordnungsblatt, June 26, 1939. The order is numbered 569 and dated June 12, 1939.

Tannenberg

By order of the Führer and Supreme Commander of the Wehrmacht, a new commemorative cuff title "Tannenberg" was to be worn by all officers, NCOs and men of "Aufklärungsgruppe 10."[11]

Kriegsberichter der Luftwaffe

By an order dated November 20, 1940, a new commemorative cuff title "Kriegsberichter der Luftwaffe" was initiated for Luftwaffe war correspondents on all fronts.[12] The first shipments were available for delivery in January of 1941. Officers had a 3 mm wide silver tresse on the top and bottom of the cuff title.

Jagdgeschwader Udet and Jagdgeschwader Mölders

By special decree of the Führer and Supreme Commander of the Wehrmacht, two new cuff titles were initiated. All officers, NCOs and men of "Jagdgeschwader 3" were now to wear the commemorative cuff title "Jagdgeschwader Udet," and all officers, NCOs and men of "Jagdeschwader 51" were now to wear the commemorative cuff title "Jagdgeschwader Mölders."[13]

[11] Luftwaffen-Verordnungsblatt, September 18, 1939. The order is numbered 813 and dated September 5, 1939.

[12] Luftwaffen-Verordnungsblatt, December, 1940. The order is numbered 1518 and dated November 20, 1940.

[13] Luftwaffen-Verordnungsblatt, January 12, 1942. The order is numbered 59 and dated December 20, 1941.

A Feldwebel in "Jagdgeschwader Mölders" (note cuff title)

Afrika (Theater of Operations)

On February 25, 1942, Reichsmarschall Göring declared that all members of the Luftwaffe stationed in Africa were to wear a dark blue cuff title "Afrika." Regulations for the wearing of this cuff title were as follows:

1. The cuff title was worn:

   (a) by all members of units and formations stationed in North Africa. Those who merely flew to or over Africa on missions were not classified as "stationed in Africa."

   (b) by those who had been wounded in Africa and transferred to military hospitals in Europe.

   (c) by those who are stationed in Africa but on furlough in Europe. In these cases, the cuff title may be worn on the blue-grey uniform.

2. The cuff title was removed when units were transferred out of Africa or when individuals were transferred to units not stationed in Africa.

3. The cuff title was 3.3 cm wide, and in dark-blue material with the word "Afrika" in aluminum embroidery for officers and officials of officer rank. "Afrika" was in matte-grey embroidery for NCOs and men. It was worn approximately 16 cm from the bottom of the right sleeve of the tropical tunic and above the cuff of the blue-grey service tunic.[14]

[14] Luftwaffen-Verordnungsblatt, March 16, 1942. The order is numbered 622 and dated March 6, 1942.

Kreta (Campaign Decoration)

Shortly after the fall of Crete, Hitler ordered the production of a cuff title "Kreta" as a campaign decoration. Its base was white with the word "Kreta" between two palm emblems and a 1 cm wide border on top and bottom in golden-yellow thread. It was to be worn above the left cuff of the tunic and great-coat and in the same approximate position for cuffless tropical tunics.

Since the cuff title was a campaign decoration, it was accompanied by an award document (see illustration of document for parachutists). It was available to all branches of the service whose members had contributed to the battle, either on the ground, in the air, or at sea. It was authorized for the following"

(a) soldiers who had pararchuted or landed by glider (including crews) on Crete up to May 27, 1941 (to this date, enemy resistance was still strong),
(b) flying units who had engaged in the battle over Crete or over the sea around the island (such as fighter, reconnaissance, bomber, Stuka and ground-support elements) before May 27, 1941,
(c) and also naval units who had engaged in combat in the waters around Crete on May 19, 1941.

After the battle, special thanks was extended to the commander of Luftlotte 4, Generaloberst Löhr, for his participating units (parts of XI. Fliegerkorps), and to the commanding general of XI. Fliegerkorps, General der Flieger Student, for all participating units under his command.[15]

NOTE:
The date of termination for awarding the "Kreta" cuff title was October 31, 1944.[16]

[15] Luftwaffen-Verordnungsblatt, October 5, 1942. The order was numbered 2519 and dated September 29, 1942.

[16] Luftwaffen-Verordnungsblatt, September 25, 1944. The order was numbered 1315 and dated September 15, 1944.

KRETA

MALEMES CHANIA RETHIMNON IRAKLION

IM NAMEN DES FÜHRERS
UND OBERSTEN BEFEHLSHABERS
DER WEHRMACHT
VERLEIHE ICH
DAS

ÄRMELBAND »KRETA«

Gefr. Karl Heinz

O; U; , DEN 20. Mai 1943

DER KOMMANDIERENDE GENERAL
DES XI. FLIEGER-KORPS

Student

GENERAL DER FLIEGER

"Kreta" award document for parachutists signed by General der Flieger student.

Afrika (Campaign Decoration)

In February 1943, the Führer ordered that the "Afrika" cuff title now be available to all three branches of the Wehrmacht as an official campaign decoration. This replaced all other "Afrika" and "Afrikakorps" cuff titles.

The cuff title was 3.3 cm wide, of camel hair and in khaki-brown. The word "Afrika," on the cuff title, was flanked by two palm trees (all in silver colored cotton thread) and edged on top and bottom by a silver-colored tresse. It was worn 1 cm above the left cuff of the tunic and great-coat, and in the same approximate position for cuffless tropical uniforms.

Besitzurkunde

Ich verleihe dem

Prüfmeister

Alfred A d e

das

Ärmelband

- Afrika -

Hauptquartier des O.B.S., den 1.5.1943

Der Oberbefehlshaber Süd

Keßelring

Generalfeldmarschall

F. d. R.:

Major

The "Afrika" award document

The awarding was based on the following regulations:

(a) a minimum of six months on African soil. Again, those who landed with supplies, etc., or merely flew over the country were not classified as "stationed on African soil."

(b) being wounded in combat in the North African theater (on land or in the air).

(c) contraction of an illness while in the North African theater of the war, which demanded evacuation to the continent. Service was required for at least three months before contraction of the illness, under these circumstances.

Any of the above requirements were considered fulfilled if an individual was killed while on duty in North Africa. In this case, only the award document was presented to his dependents.[17]

NOTES:

1. The Führer later decreed that from May 6, 1943, those who fought in the final phases of the African campaign will have the six month service time reduced to four months.[18]

2. Those who fought in the African campaign and were awarded the Iron Cross, German Cross in gold, Goblet of Honor or the Salver of Honor, were now entitled to the "Afrika" cuff title regardless of service time.[19]

3. For members of the Luftwaffe, the cuff title "Afrika" was no longer to be awarded after August 28, 1944, with the exception of prisoners of war, those missing-in-action and those confined due to infectious diseases who could possibly be returned to active duty.[20]

Additional cuff titles worn by Luftwaffe personnel

"Führer-Hauptquartier" - (Führer's Headquarters) Luftwaffe personnel assigned to Hitler's headquarters wore the army's "Führer-Hauptquartier" cuff title (gold on black) on the lower left sleeve. During the war, this cuff title was changed to silver on black.

"Führerbegleitbattalion" - (Führer Escort Battalion) Luftwaffe personnel who manned this Flak battalion, which was assigned to the "Führerbegleitbrigade," wore the Army's "Führerbegleitbattailon" cuff title on the lower left sleeve.

"Düsenjäger-Versuchskommando 1/A" - (Jet Fighter Experimental Detachment) This cuff title was initiated in early spring 1945, and was blue with aluminum letters and edging.

[17] Luftwaffen-Verordnungsblatt, February 8, 1943. The order was numbered 283 and dated January 1943.

[18] Luftwaffen-Verordnungsblatt, May 31, 1943. The order was numbered 1028 and dated May 20, 1943.

[19] Luftwaffen-Verordnungsblatt, January 31, 1944. The order was numbered 132 and dated January 12, 1944.

[20] Luftwaffen-Verordnungsblatt, September 4, 1944. The order was numbered 1180 and dated August 27, 1944.

NOTE:

The following cuff titles do exist but no documentation has been uncovered to authenticate them.

"Galland" - Presumably worn by members of "Jagdverband (JV 44 Galland" (formed in early 1945), which was comprised of the Me 262 jet fighter and led by Generalleutnant Galland himself. The cuff title was blue with aluminum lettering and edging.

"Geschwader Lützow" - It is believed that this cuff title is from a pre-war propaganda movie concerning a fictitious "Geschwader Lützow."

"Turbinenjäger-Versuchskommando I"

"Turbinenjäger-Versuchskommando II"

} Turbo-jet fighter experimental detachments 1 & 2

NOTE:

All Fallschirmjäger cuff titles and those from Regt. "General Göring" and Brigade/Division "Hermann Göring" are described and illustrated in their appropriate section of this book.

Armbands

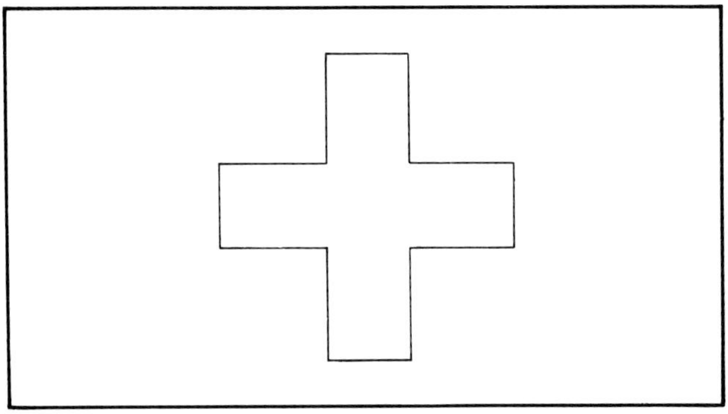

International Red Cross armband
(red on white)

NOTE:

Worn on the upper left sleeve of the service tunic, flying blouse and great-coat.[21]

Auxiliary stretcher bearer armband (black on white)

NOTE:

Worn on the upper right sleeve of the service tunic, flying blouse and great-coat.[22]

[21] Luftwaffen-Verordnungsblatt, January 10, 1938. The order number is 23 and dated January 5, 1938.

[22] Ibid.

Maneuver umpires/airfield traffic director's armband (white)

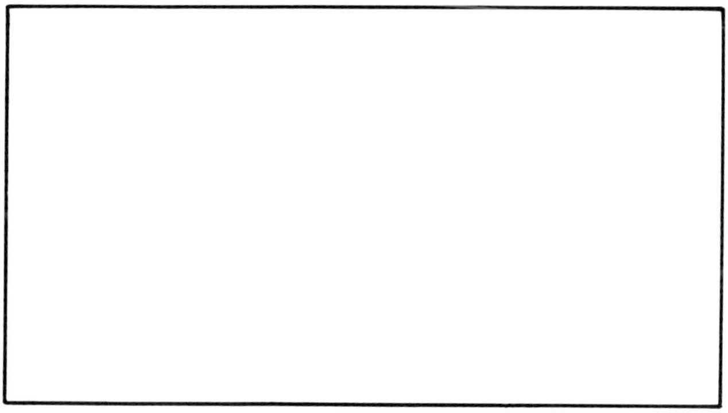

This white armband was worn on the upper left sleeve (Maneuver Umpires) of the service tunic, flying blouse and great-coat. A white cap band was worn by Airfield Traffic Directors and Maneuver Umpires.[23]

*worn on the upper right sleeve by Maneuver Observers.

Real Estate assessing personnel armband

This plain green armband was worn on the upper left sleeve of the service tunic, flying blouse and great-coat.

Armband for civilian personnel in Luftwaffe auxiliaries (blue on white)

The blue on white "Deutsche Luftwaffe" armband was worn on the upper left sleeve extensively by civilian personnel functioning within a Luftwaffe-sponsored auxiliary organization, such as the Heimatflak.

[23] Luftwaffen-Verordnungsblatt, February 7, 1938. The order is numbered 99 and dated February 3, 1938.

Armband for Luftwaffe personnel assigned to fire fighting duty (light blue)

A plain, light blue, 13 cm wide armband was worn on each upper sleeve of their service tunic, flying blouse and great-coat. Each armband was marked with the appropriate garrison headquarters stamp.

Armband for officers and personnel assigned to railway station security

Luftwaffe officers and personnel assigned to railway station security and duty were authorized to wear on their upper left sleeve, a plain, white, linen armband (with the appropriate issuing office hand stamp.) The commanding officer of a railway station detachment wore the white armband also, but with "Bahnhofskommandantur" in black letters on it. As of the order date, January 19, 1937, this armband replaced a yellow armband, of the same specifications, which was previously worn for this duty.[24]

A young parachutist on railway security duty in Italy.

Armband for qualified Luftwaffe interpreters (black on pink)

The armband for qualified interpreters in the Wehrmacht was of pink linen, 7 cm wide and had "Wehrmachtdolmetscher" in black letters on it. It was worn on the upper left sleeve. Only after extensive schooling and testing was this armband to be given to qualified interpreters, and then only by the head of the training facility.[25]

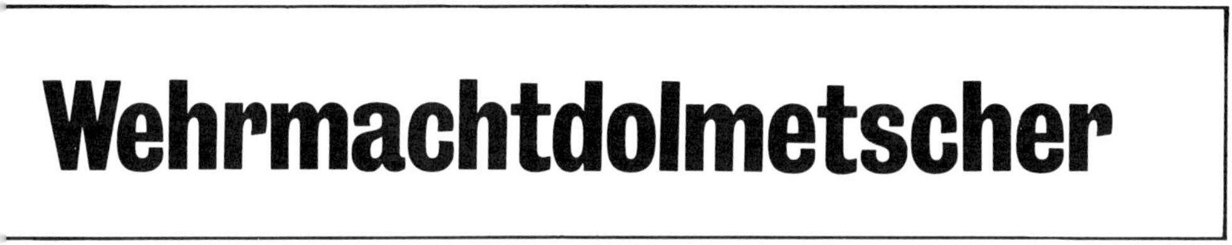

Armband for flying crews (black on yellow)

With the advent of enemy troops on German soil, a new method of marking downed flying personnel, who were making their way back to their lines was required. A number of incidents had occurred in which such personnel had been mistaken for advancing enemy troops and killed. The introduction and wearing of this yellow armband with black Luftwaffe eagle and the words "Deutsche Luftwaffe" was an attempt to solve this problem.[26]

[24] Luftwaffen-Verordnungsblatt, February 1, 1937. The order is numbered 109 and dated January 19, 1937.

[25] Luftwaffen-Verordnungsblatt, January 18, 1943. The order is numbered 118 and dated January 11, 1943.

[26] Luftwaffen-Verordnungsblatt, January 8, 1945. This late order is numbered 26 and dated December 30, 1944.

Special Uniforms and Insignia

Uniform of the Reichsmarschall des Grossdeutschen Reiches

The first pattern

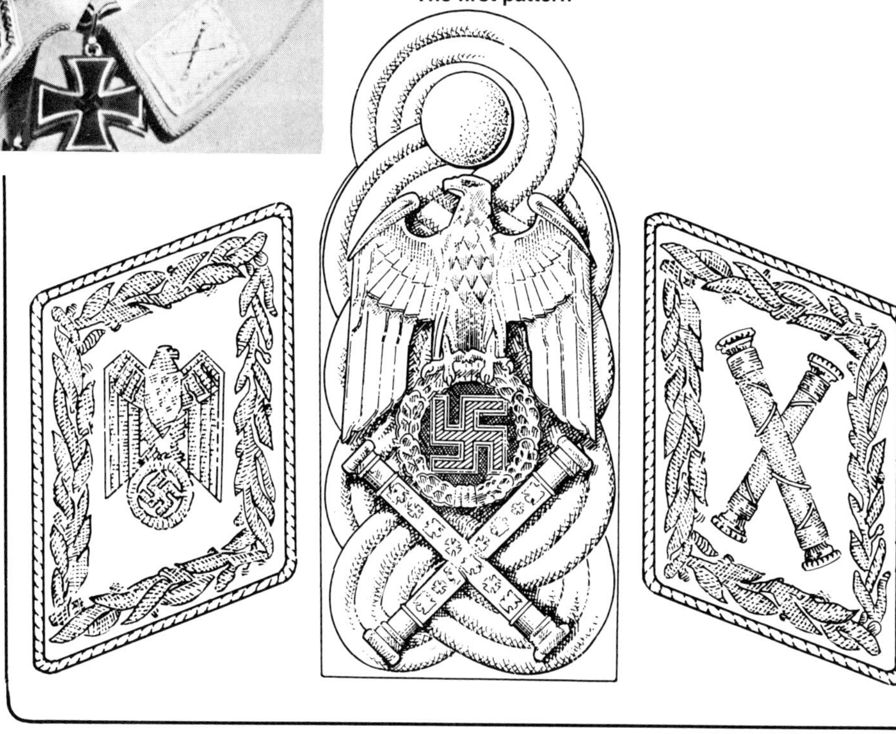

On July 19, 1940, the Führer designated Hermann Göring "Reichsmarschall of the Great German Reich" and awarded him the Grand Cross of the Iron Cross. Therefore, in August of 1940 a special Reichsmarschall uniform and insignia was introduced for wear. The first pattern rank insignia consisted of gold cord shoulder straps on a white base, with a gold Wehrmacht eagle over crossed Field Marshal's batons. The collar patches were of a silver brocade base with a matte gold piping. On the right collar patch was a gold hand-embroidered

Reichsmarschall Göring with chief of the Luftwaffe General Staff, General der Flieger Jeschonnek.

The second pattern

Wehrmacht eagle surrounded by gold hand-embroidered laurel leaves. On the left collar patch were gold hand-embroidered crossed Field Marshal's batons surrounded by gold hand-embroidered laurel leaves. This first pattern set of collar patches was worn until March of 1941. After this date, the second pattern set (crossed batons on both collar patches) was worn with the same style shoulder straps until the end of the war.

The Reichsmarschall uniform was in the same cut as that of a Luftwaffe Generalfeldmarschall, but with special rank insignia and in pigeon-grey. The tunic, which carried the Reichsmarschall rank insignia, had the collar piped in gold cord, the breast eagle in gold hand-embroidery on a white base and a single row of hidden buttons (see photo below). The Reichsmarschall also wore a tunic similar in design to the special tunic for generals, but in pigeon-grey, with gold piping around the collar, white piping around the lapels and cuffs (no white lapels) and the Reichsmarschall insignia.* It also had two rows of three gold buttons and a gold, hand-embroidered breast eagle. Wide, white stripes were worn on the pigeon-grey straight pants only, whereas white piping was worn on the breeches (for a short time, however, white stripes were also worn on the breeches). The boots and belt for this uniform were in a reddish-brown leather.

*This, plus the previously described tunic, were also in white for summer wear.

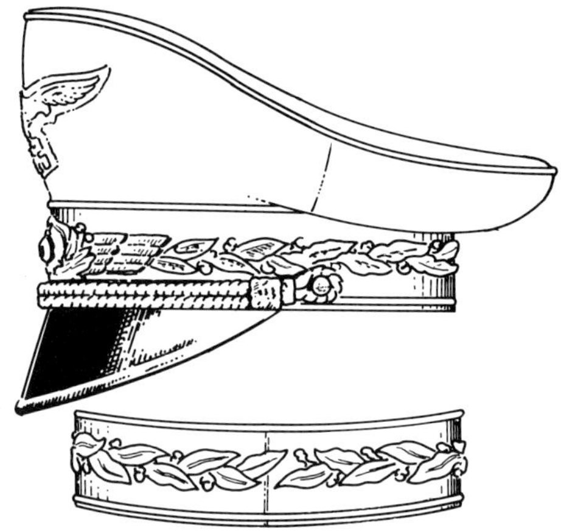

NOTE:
Göring surrendered in 1945 wearing a tunic with a single row of five buttons, breast and side pockets, no breast eagle or collar patches, but with shoulder straps.

The peaked cap was also in pigeon-grey with the following distinctions. The Luftwaffe eagle was larger than normal and in gold wire embroidery. The special winged Reich cockade, within a laurel leaf wreath, was also in gold wire embroidery. The cap was piped in gold and had gold cap cords as for general ranks. It also had a continuous laurel leaf pattern in gold embroidery going around the grey velvet cap band. The peak was in black leather.

The great-coat, also in pigeon-grey, was as that of the Luftwaffe Generalfeldmarschall with white lapels and piping, two rows of five gold buttons (top two of which were worn open), and Reichmarschall collar and shoulder rank insignia. In addition, Göring had a pigeon-grey great-coat with one row of hidden buttons, white piping around the collar, cuffs and down the front, plus the Reichsmarschall rank insignia.

Reichsmarschall eagle for the cape

A cape, in the same cut as that of the Luftwaffe, was also part of this unique uniform. It was in pigeon-grey with gold piping around the collar and with the gold clasp and chain (as that for Luftwaffe general officers). A large Reichsmarchall eagle was positioned on the upper left arm of the cape in gold hand-embroidery on a white background.

The Reichsmarschall peaked cap also existed without the oakleaf embroidery around the cap band (in white for summer wear also).

Göring's Reichsmarschall peaked cap, baton and swagger stick.

When Hitler designated Göring as "Reichsmarschall" on July 19, 1940, he also bestowed upon him the Reichsmarschall baton. It was 49 cm long and 33 cm in diameter. It was similar to a baton for a Luftwaffe Generalfeldmarschall, but with the following variations: it was of ivory with the alternate eagles, Balkenkreuze and Eiserne Kreuze in gold, both coronets (top and bottom) were in gold and adorned with diamonds, and had Göring's name plus date of bestowal on the base coronet. It was carried only on special occasions.

A special swagger stick was carried daily when the baton was not specified. It was 85 cm long, also in ivory with metal fittings in gold and platinum, and with a black/white/red twisted cord with tassel.

NOTE:
The Luftwaffe "Marschallstab" was in a light blue velvet color with gold fittings (with eagles and iron crosses). The Feldmarschall swagger stick was black with gold fittings.

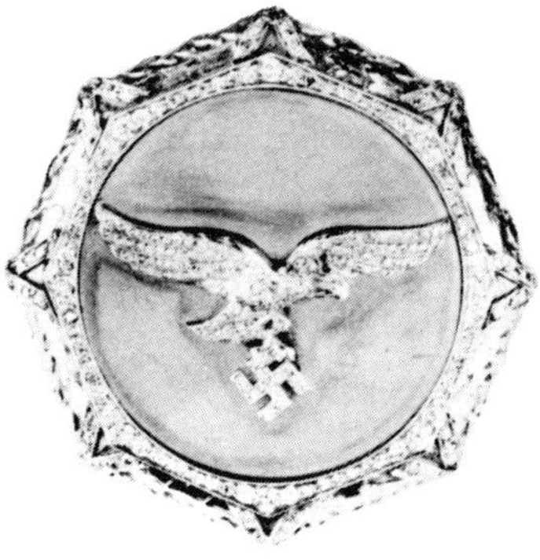

Top coronet

Bottom coronet

The Reichsmarschall baton

A special Reichsmarschall dagger was produced by students of the Berlin Technical Academy and presented to Göring later in 1940. The fittings and scabbard were finished in gold, the pommel was adorned with diamonds and rubies and the fluted grip was of ivory. The cross-guard carried the spread-winged Reichsmarchall emblem on its face and had an etched Iron Cross on each of its arms (upper surface). The scabbard was also fluted its full length with an oakleaf pattern at its base. The dagger was suspended by a specially designed double-strap hanger (see illustration).

The dress belt was in gold brocade with two black stripes and one red stripe in the center. The buckle was the same design as for Luftwaffe officers but in gold, including the Luftwaffe eagle. The standard leather belt with double-claw buckle, was in a reddish-brown leather with gold fittings.

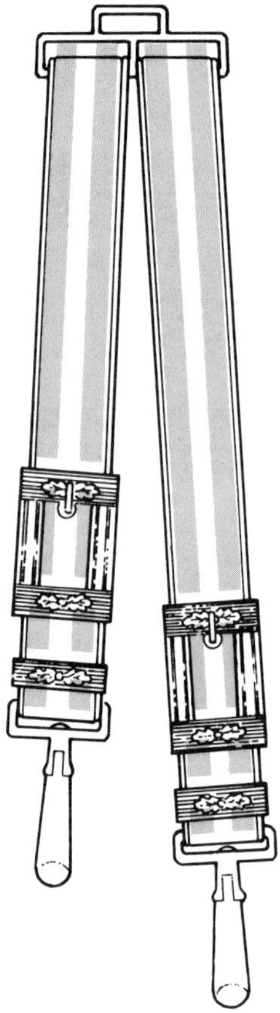

Regiment "General Göring"

During the early thirties, police battalions were organized to protect and support Nazi leaders on political tours and at engagements. It was at this time that the Landespolizei-Gruppe "General Göring" (retitled on December 22, 1933, previously Landespolizei-Gruppe Wecke z.b.V.) was assembled as a bodyguard for Secretary of the Interior, Hermann Göring.

NOTE:
Members of Landespolizei-Gruppe "General Göring" wore the green Landespolizei uniform with a green cuff title "Landespolizeigruppe General Göring" on the lower left sleeve.

The Gruppe served in this security capacity until September 23, 1935, when it was officially transferred to the newly formed Luftwaffe as Regiment "General Göring." For six months, the Regiment's members continued to wear their Landespolizei uniforms with the Luftwaffe breast eagle and a green cuff title "General Göring" (lower right sleeve). On March 23, 1936, a major uniform change was ordered for the Regiment. Members of Regiment "General Göring" now wore the standard Luftwaffe uniform but with white Waffenfarbe. As a further regimental identification, its personnel wore a 3.3 cm wide, blue cuff title "General Göring" on the lower right sleeve. Its specifications were as follows:[1]

(a) officers — "General Göring" in aluminum wire embroidery and the upper and lower trim in aluminum braid
(b) NCOs — "General Göring" in matte-grey embroidery and the upper and lower trim in matte-grey braid
(c) other ranks — "General Göring" in matt-grey embroidery and no trim

"General Göring" cuff title for officers

The Regiment as of December 1, 1935, consisted of:
Regimentstab
I. Jäger-Bataillon
II. Jäger-Bataillon
III. Flak-Abteilung

The Waffenfarbe for the above units was white, but with the following additions:[2]
(a) officers - white with aluminum embroidery and aluminum piping
(b) NCOs and men (except those from the Flak-Abteilung) - white with rifle-green piping
(c) NCOs and men (from the Flak-Abteilung) - white with bright red piping

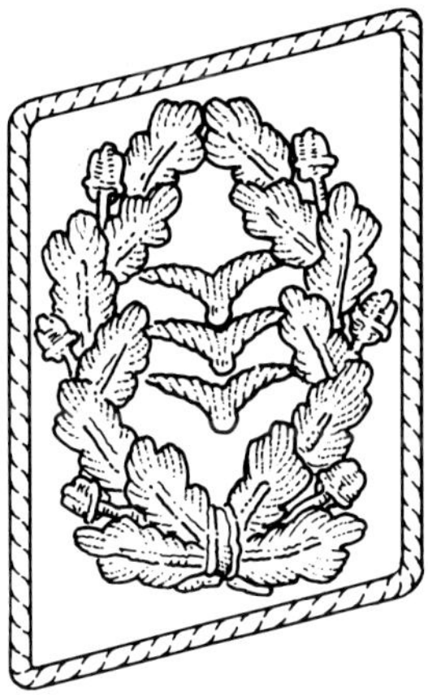

Officer's collar patch

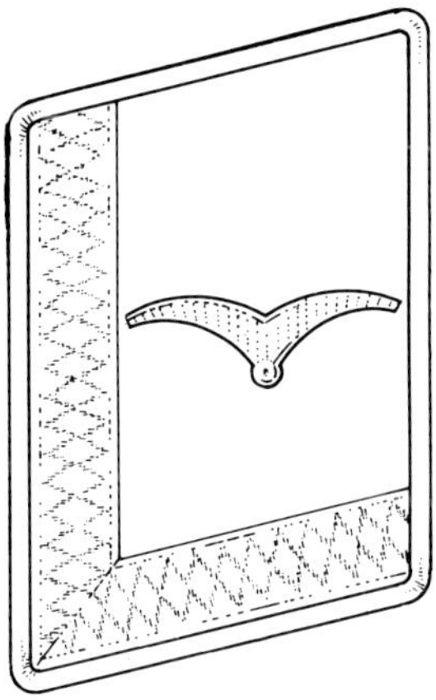

NOC's collar patch

In addition to the three battalions, a Wach-Kompanie (guard company) was organized on July 11, 1936, with a second company being formed on April 1, 1937. These were incorporated into the new regimental composition on October 1, 1937.

Regimentstab
I. schwere Flak-Abteilung, (8.8 cm heavy Flak battalion)
II. leichte Flak-Abteilung, (2 cm light Flak battalion)
III. Wach-Bataillon (guard battalion)
IV. Fallschirmschützen-Bataillon (paratroop battalion)

On April 1, 1938, the IV. Fallschirmschützen-Bataillon was removed from the Regiment and stationed at Stendal where it became the I. Bataillon/Fallschirmjäger-Regiment 1. Its members would soon become the training personnel for Parachute School Stendal. By November 1, the Regiment had been expanded and now consisted of:

Regimentstab (with a signals platoon and a music corps)
I. schwere Flak-Abteilung
II. leichte Flak-Abteilung
III. Scheinwerfer-Abteilung (searchlight battalion)
IV. leichte Flak-Abteilung
Wach-Bataillon

The Regiment was an elite unit and proudly participated in the Vienna parade on March 15, 1938, after the Austrian Anschluss, in the occupation of the Sudetenland and in the final take-over of the remainder of Czechoslovakia. It remained in Germany until the spring of 1940. At this time, elements of the Regiment saw their first combat in the Norwegian campaign. The bulk of the Regiment, however, took part in the Belgian campaign on May 10, 1940, and later

[1] Luftwaffen-Verordnungsblatt, March 23, 1936. The order was numbered 348 and dated March 12, 1936.

[2] Luftwaffen-Verordnungsblatt, April 13, 1936. The order was numbered 453 and dated April 3, 1936.

in the invasion of Holland. After the fall of France, the Regiment was assigned to antiaircraft duty near Paris for a short time. Later that year, it was reassigned to the Reich. The Regiment remained basically dormant until June 1941, when many elements of the Regiment engaged in the early stages of the German invasion of Russia. During the spring of 1942, the Regiment was enlarged to brigade strength and was officially titled "Brigade Hermann Göring." The "General Göring" cuff title continued to be worn until late August of 1942.

Wachbataillon der Luftwaffe Berlin

The Wachbataillon der Luftwaffe Berlin was stationed in Berlin where it served as security (as well as supplying messengers) for the Reich Air Ministry and at the residence of the Reich-Minister of Civil Aviation. The Wachbataillon consisted of a staff and two Schützenkompanien (infantry companies), of three platoons each.

The Wachbataillon personnel wore the golden-yellow Waffenfarbe. The permanent cadre of this unit wore a "LW"* on their shoulder straps and transferred personnel into the unit wore on the base of their shoulder straps, a 0.5 cm cord in Waffenfarbe.[3]

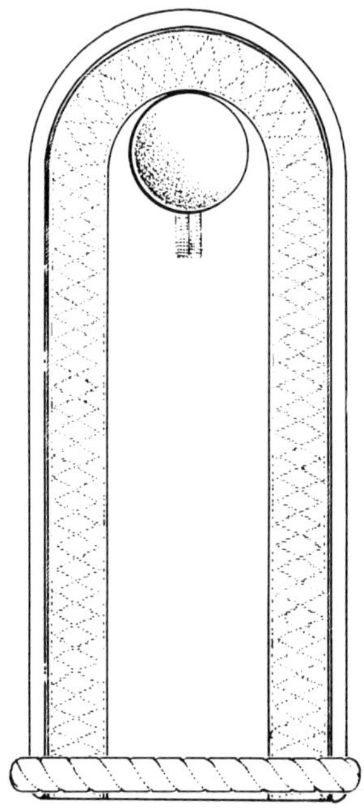

*officers - gilt colored metal

NCOs to Oberfeldwebel - aluminum colored metal

Unterfeldwebel and below - golden-yellow embroidery

NOTE:
Later in the war, the Wachbataillon became the 1. Schützen-Bataillon, and then the Stabs-Bataillon RLM.

[3] Eberhard Hettler, *Uniformen der Deutschen Wehrmacht*, p. 96.

Division "Hermann Göring"

When Brigade "Hermann Göring" was formed in early 1942, it consisted of one Schützen-Regiment and one Flak-Regiment.* An order dated May 22, 1942, stated that the "General Göring" cuff title should be completely replaced by the new "Hermann Göring" cuff title no later than late August of that year.[4] In June, the Brigade was enlarged to divisional strength and titled Division "Hermann Göring." The first pattern "Hermann Göring" cuff title was in gothic script but was shortly replaced by a cuff title in block letters.

*The Brigade title evolved from Regiment (mot) "Hermann Göring," which was previously Regiment "General Göring."

Gothic pattern (worn also by Brigade members)

Block pattern

By an order dated January 4, 1943, the Division underwent an extensive rank insignia change. General officers, general staff officers, officials, medical officers and other medical personnel were not affected by this order. Neither was the field postal service which utilized Army uniforms. Officers and all other ranks continued to wear collar patches with white Waffenfarbe (officers with silver piping), but individual branch color piping was now required and was as follows:[5]

Grenadierregimenter Wachregiment	white piping (small black line between the piping and the collar patch)
Panzerregiment Panzerjägerabteilung Aufklärungsabteilung	pink piping

[4] Luftwaffen-Verordnungsblatt, June 1, 1942. The order was numbered 1381 and dated May 22, 1942.

[5] Luftwaffen-Verordnungsblatt, January 18, 1943. The order was numbered 127 and dated January 4, 1943.

Artillerieregiment
Flakregiment } red piping
Führer-Flakabteilung
Nachrichtenabteilung - golden-brown piping
Pionierbataillon - black piping
Nachschubtruppen
Verwaltungstruppen } light-blue piping
Feldgendarmerie

NOTE:
Shoulder strap underlay and piping was still white as before.

A further order concerning collar patches and shoulder straps was issued on April 2, 1943. Collar patches were still in white but from this date no branch piping was worn (officers retained the silver piping). The shoulder straps were now to carry the branch color piping.

Divisionsstab - pink piping
Grenadierregimenter
Wachregiment* } white piping
Jägerregiment - rifle green piping
Panzerregiment
Panzerjägerabteilung } pink piping
Panzeraufklärungsabteilung
Flugbereitschaft } golden-yellow piping
Artillerieregiment
Flakabteilung } red piping
Führer-Flakabteilung*
Nachrichtenabteilung - golden-brown piping
Pionierbataillon - black piping
Nachschubtruppen
Verwaltungstruppen } light-blue piping
Feldgendarmerie

*Only these two units were permitted to wear collar patches on their great-coats at this time.

NOTE:
The Feldgendarmerie branch color was changed to orange by an order dated June 3, 1943.[6]

[6] Lufftwaffen-Verordnungsblatt, June 15, 1943. The order was numbered 1115 and dated June 3, 1943.

Collar insignia for the Panzerregiment and Panzerjägerabteilung was the Panzer collar patch with skull, but in white (note shape).

Waffenfarbe by an order dated April 2, 1943.

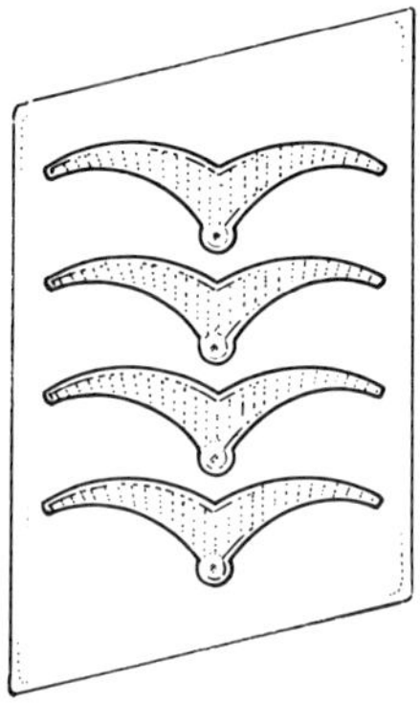

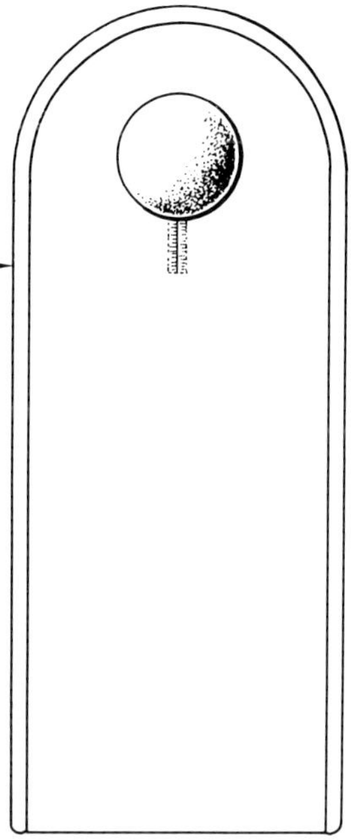

(Above) Göring greets the men of his division. Standard collar patches with white Waffenfarbe are worn.

(Right) Note Hauptmann Rossmann, a Knight's Cross holder, is wearing the black army Panzer uniform with army-style collar patches (black piped in white). The division utilized army insignia for a time after its formation.

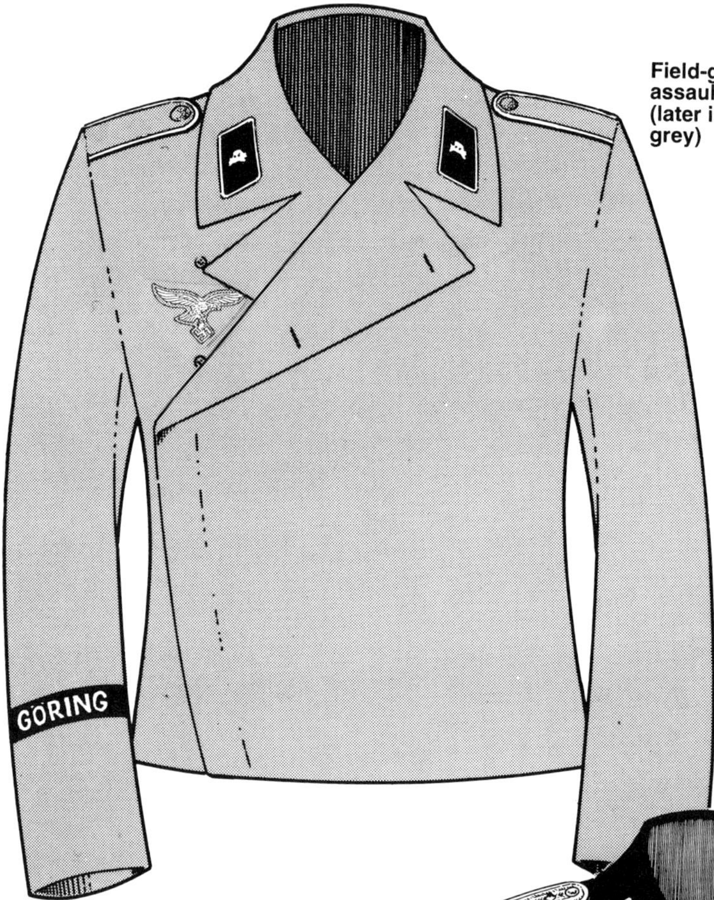

Field-grey tunic for assault gun crews (later in Luftwaffe blue-grey)

Black tunic for tank and armored car crews (the black uniform, with black collar patches piped in white, plus Panzer beret was initially worn by the Panzer-Späh-zug in 1938).

The Parachute Troops

On April 1, 1938, the IV. (Fallschirmschützen-) Bataillon was transferred from Regiment "General Göring" to Stendal where it became the Luftwaffe's I. Bataillon/Fallschirmjäger-Regiment 1. On January 1, 1939, the Army's Fallschirm-Infanterie-Bataillon was also absorbed into the Luftwaffe's budding paratrooper force where it became the II. Bataillon/ Fallschirmjäger-Regiment 1.* These Battalions were the nucleus of General Kurt Student's 7. Flieger-(Fallschirm-) Division.

*The Fallschirm-Infanterie-Bataillon, initially under the command of Major Richard Heidrich, came under Luftwaffe control after insistance from Hermann Göring that all parachute units should be under one central command. The Bataillon, when absorbed into the Luftwaffe, was permitted to retain its infantry, swallow-tailed standard, for traditional purposes.

Standard for I. Bataillon/ Fallschirmjäger-Regiment 1

The light green Fallschirmjäger-Regt. 1 cuff title was instituted on August 19, 1939, and worn by all members of that unit on the lower right sleeve. Its specifications for officers, NCOs and men were as those for the Regiment "General Göring" cuff title. When the IV. (Fallschirmschützen-) Bataillon was transferred from Regiment "General Göring," it was permitted to retain its standard (identical to that of the I. Bataillon/Regiment "General Göring" but without the Kolonialpolizei badge directly below the wreath).

Fallschirm-jäger Rgt. 1 cuff title

Fallschirm-jäger Rgt. 2 cuff title

The "Fallschirm-Division" cuff title, in dark green with aluminum wire embroidery for officers and in aluminum colored thread for other ranks, was for paratroopers and all other remaining troops of 7. Flieger- (Fallschirm-) Division (including the Fallschirmschule), with the exceptions of the Fallschirmjäger-Regt. 1 and 2. These Regiments retained their own cuff titles. None of the three official paratroop cuff titles were worn after the outbreak of war in 1939.

NOTE:
The "Fallschirm-Division" cuff title did not have upper and lower trim for officers and NCOs, as did the Regiment "General Göring" cuff title.

Fallschirm-Division cuff title for the 7. Flieger-(Fallschirm) division

The paratrooper* wore the standard Luftwaffe issue blue-grey uniform with golden-yellow Waffenfarbe. The blue-grey bloused pants were loose-fitting with a buttoned side pocket on the right leg (for the utility knife, etc.). A special jump boot was also worn which laced up the side. Shortly after the Crete operation, however, a new issue jump boot was introduced which laced at the front.

* All branches (Waffengattungen) of all divisions

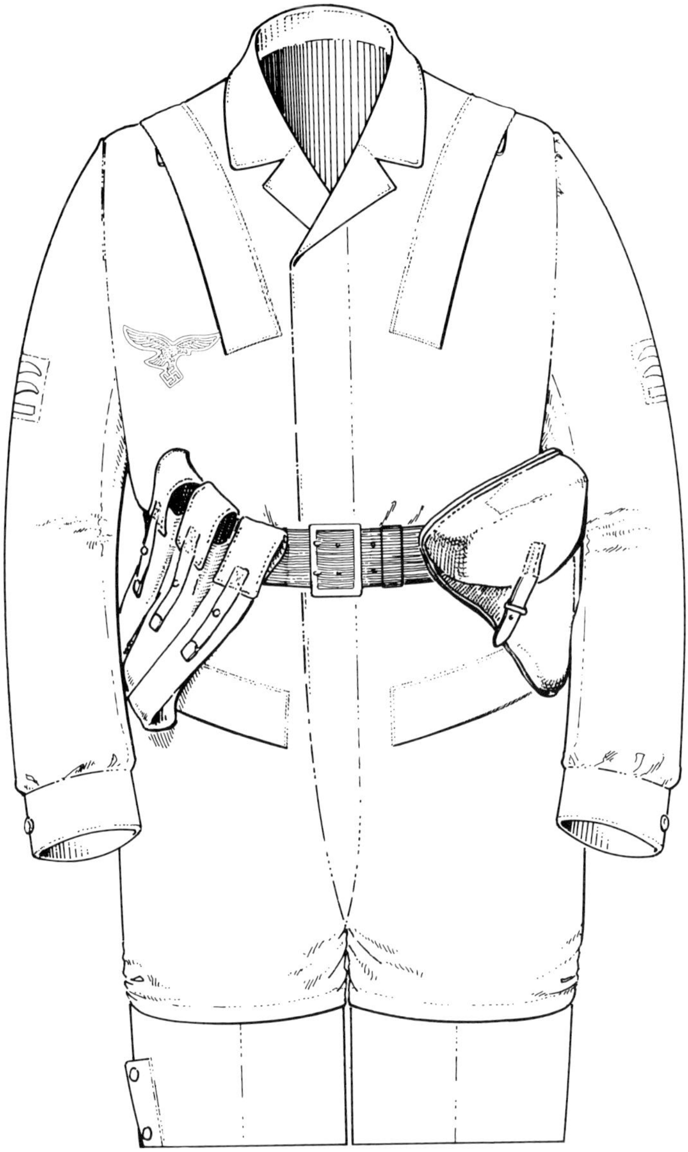

Oberleutnant's jump smock with MP40 clip pouches and P38 holster (2nd model jump smock)

NOTE:

The jump smock, which was worn over the standard tunic and bloused pants and of a waterproof gabardine fabric, was fitted with a zipper or button fly-front for rapid disrobing. Two large fly-fronted pockets were on the upper chest, running diagonally and two large fly-fronted pockets on the thighs, all of which were fastened with course plastic zippers.

The paratrooper's most distinctive article of clothing was the jump smock. It was initially a step-in model in either blue-grey or olive-green. The Luftwaffe breast eagle was worn on the right breast. By the spring of 1940, a camouflage pattern jump smock had been introduced...in a "splinter" pattern. Shortly before the Crete invasion, a new variation jump smock was introduced. It could be buttoned at the crotch during the jump, and loosened upon landing, giving the paratrooper a greater freedom of movement. Shortly after the Crete battle, another camouflage pattern was also utilized...a "splotched" pattern. The two camouflage patterns were predominantly worn for the duration of the war. Paratroopers stationed in the Mediterranean area (Italy, Sicily, Africa, etc.) wore a tropical tunic, shirt and bloused pants with a high buttoned pocket on the left leg. A tan jump smock was also issued to some units.

The "splinter" pattern

The "splotched" pattern (note SS belt buckle)

Special protective items were also worn during the jump. Rubber padded protectors were strapped onto the knees and elbows and a long gauntlet-type leather pair of gloves was worn. These were all discarded after the jump, when the situation permitted.

NOTES:

1. The special paratrooper helmet is covered in the headgear section of this book.
2, Rank insignia utilized on the jump smock was the double wing and bar system, on the appropriate colored backing.

Rubber padded knee protectors fastened in place before the jump on Crete.

Riflemen wore a 100-round, multi-compartment bandoleer around their neck, whose ends were attached to the waist belt. The bandoleer was issued in blue-grey, olive-green, and in the "splotched" and "splinter" camouflage patterns.

Man in left center is wearing an eight-compartment bandoleer for the Fallschirmjäger-Gewehr (FG42). Man at far right is holding the MG42.

Paratrooper's bandoleer

Field Divisions

Luftwaffe Field Divisions were formed in late 1942 to replace the Army's high losses in men on the Russian front. This new source of personnel was drawn from surplus personnel of Flak and signals troops, ground crews of the flying troops and administrative units. They were later expanded with recruits, both German and foreign. The Divisions consisted of two Lw. infantry regiments, with three battalions each, plus supporting units (an exception was the 21. Lw.-Felddivision which had three regiments, with three battalions each).

Most of the 22 divisions formed were sent to the Russian front where many of them were decimated within the next year of intense fighting. Thus, most were disbanded in late 1943 or at the latest by the summer of 1944. By a Führer order dated September 20, 1943, all remaining Felddivisionen (Luftwaffe) were to be incorporated into the Army by October 31, 1943. They were then reorganized along the lines of Army infantry divisions "1944-Type" and returned to the front to fight side-by-side with Army formations.

The uniform worn by Field Division personnel was standard Luftwaffe and their weapons and equipment was as for a normal infantry division. Initially, the Waffenfarben worn by individuals was the same as their previous branch...yellow (flying troops), red (Flak), brown (signals), blue (medical), etc. By an order dated January 4, 1943, however, a more uniform system of identifying Luftwaffe Field Divisions was established.[7]

Officers of Luftwaffen-Feldkorps wore rifle green collar patches with aluminum piping and matte-grey shoulder straps with the underlay in their own branch color (general officers were as for Division "Hermann Göring). Other ranks wore rifle green collar patches, piped in their own branch color plus shoulder straps also piped in the branch color.

Jägerregimenter, Wachkompanien } rifle-green (small black line between the piping and the collar patch)
Radfahrkompanien, Aufklärungskompanien } golden-yellow
Artillerieregimenter, Flakabteilungen } red
Panzerjägerabteilungen - pink
Nachrichtenabteilungen - golden brown

[7] Luftwaffen-Verordnungsblatt, January 18, 1943. The order was numbered 127 and dated January 4, 1943.

Pionierbataillone - black

Nachschubtruppen }
Verwaltungstruppen } light blue
Feldgendarmerie }

NOTE:
Collar patches were often not worn and the aluminum wings were positioned directly on the collar.

NOTES:

1. All 22 Field Divisions followed the above regulations with the exception of the 20th.

2. By an order dated June 3, 1943, Feldgendarmerie personnel now wore rifle-green collar patches with orange piping. Officers and other ranks wore shoulder straps also piped in orange or with an orange underlay.[8]

In January of 1943, the lowest rank of personnel in Luftwaffen-Feldkorps and Luftwaffen-Felddivisionen were changed from standard Luftwaffe designations to the following:[9]

(a) Luftwaffe infantry and antitank units (Jäger- und Panzerjägereinheiten) - Jäger

(b) Luftwaffe artillery and Flak units (Artillerie- und Flakeinheiten) - Kanonier

(c) engineer units (Pioniereinheiten) - Pionier

Luftwaffe Grenadiers march to the allied beachhead at Nettuno (March 20, 1944)

[8] Luftwaffen-Verordnungsblatt, June 15, 1943. The order was numbered 1115 and dated June 3, 1943.

[9] Luftwaffen-Verordnungsblatt, January 18, 1943. The order was numbered 130 and dated January 12, 1943.

Luftwaffe field division personnel were also issued a 3/4 length camouflage smock, with the Luftwaffe breast eagle. It was worn with or without shoulder straps and available in the "splinter" or "splotched" pattern.

An O.K.H. decree dated July 10, 1944, ordered the following Waffenfarben changes (with Army uniforms and Army rank designations):[10]

(a) All members of Luftwaffe Jäger regiments (with exception of the 20. Lw.-Felddivision) were now to wear white Waffenfarbe, in conjunction with Army regulations. The lowest rank designation remained "Jäger" and the Unteroffiziere rank was now called "Oberjäger." All other rank designations were as those for a Grenadier-Regiment.

(b) Members of Luftwaffe Füsilier battalions (honorary designation for reconnaissance units in an infantry division) were now to wear white Waffenfarbe also. The lowest rank designations were now "Füsilier" and "Oberfüsilier" (rank as Army Oberschütze). All other rank designations were as those for Grenadier-Regimenter.

(c) The Luftwaffe infantry regiments from the 20. Luftwaffen-Felddivision wore golden-yellow Waffenfarbe, the Waffenfarbe of the cavalry. All rank designations were as those for "Reiter-Regimenter. The 20. Luftwaffen-Felddivision was a cyclist division.

[10] Egon Denzel, *Die Luftwaffen-Felddivisionen 1942-1945*, p. 46.

Music Corps

Soldiers with officers' rank of the Luftwaffe's Music Corps wore the standard Luftwaffe officers' uniform with the following insignia variations (after 12.4.38).

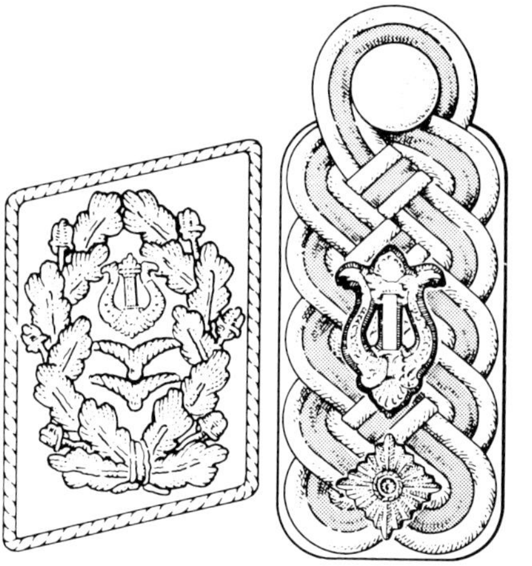

Obermusikinspizient (Oberstleutnant)

Musikinspizient (major)

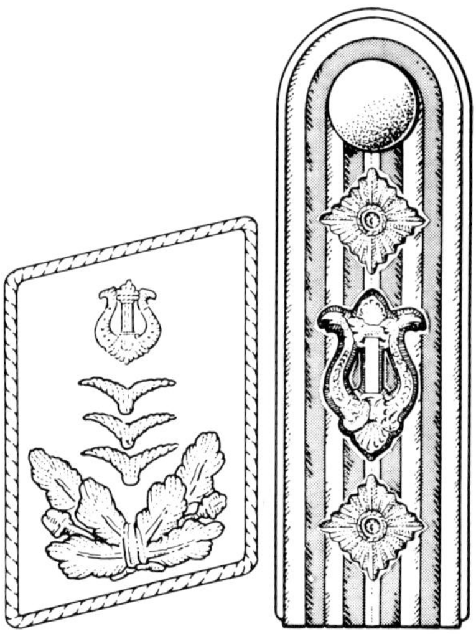

Stabsmusikmeister (Hauptmann)

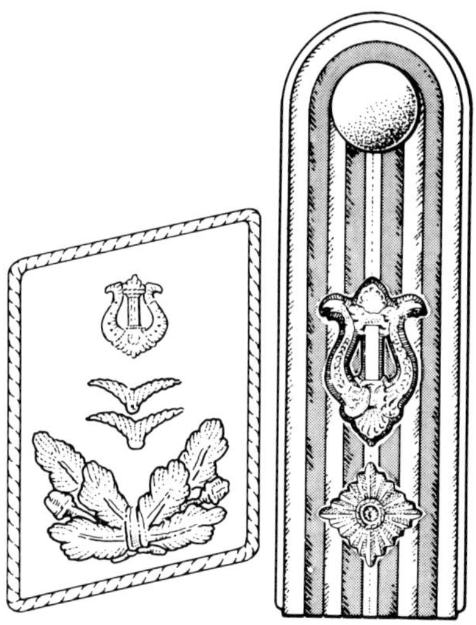

Obermusikmeister (Oberleutnant)

Musikmeister (leutnant)

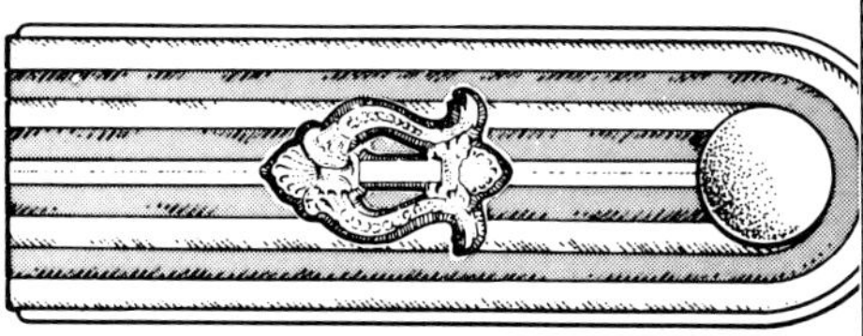

NOTES:

1. The Waffenfarbe worn by all ranks in the Music Corps was their original branch color.
2. Ranks below Musikmeister wore standard Luftwaffe rank insignia on their original branch color (Waffenfarbe).
3. Officers' shoulder straps were of alternate aluminum/red/aluminum cord, on the appropriate Waffenfarbe.
4. The ranks of Obermusikinspizient and Musikinspizient were originally on a black Waffenfarbe for they were considered Air Ministry officials. Due to the change in the RLM Waffenfarbe before the war, however, these ranks then wore their original branch color.
5. The musician's lyre and the rank pips were of a gold color for the ranks of Obermusikinspizient down to Musikmeister.
6. There was also a Musikleiter rank with Oberfeldwebel shoulder straps (aluminum lyre between pips) and plain collar patches with an aluminum lyre.

Luftwaffe musicians wore the traditional "swallows' nest" (Schwalbennester), which was detachable, on the upper two arms of the service tunic or flying blouse. There were three variations, which were as follows:

(a) Musiker (bandsmen) - the "swallows' nest" in appropriate Waffenfarbe and trimmed in aluminum braid

(b) Stabshornisten (bugle-majors) - as for Musiker but with a 7 cm long aluminum fringe

(c) Spielleute (fife and drum corpsmen) - as for Musiker but trimmed in grey braid

NOTES:

1. The Musiker from Regiment "General Göring" wore the "swallows' nest" as above but with a 4.5 cm long aluminum fringe (white Waffenfarbe).
2. The Spielleute from Regiment "General Göring" wore the "swallows' nest" as above but with a 4.5 cm long, blue-grey, wool fringe (white Waffenfarbe).

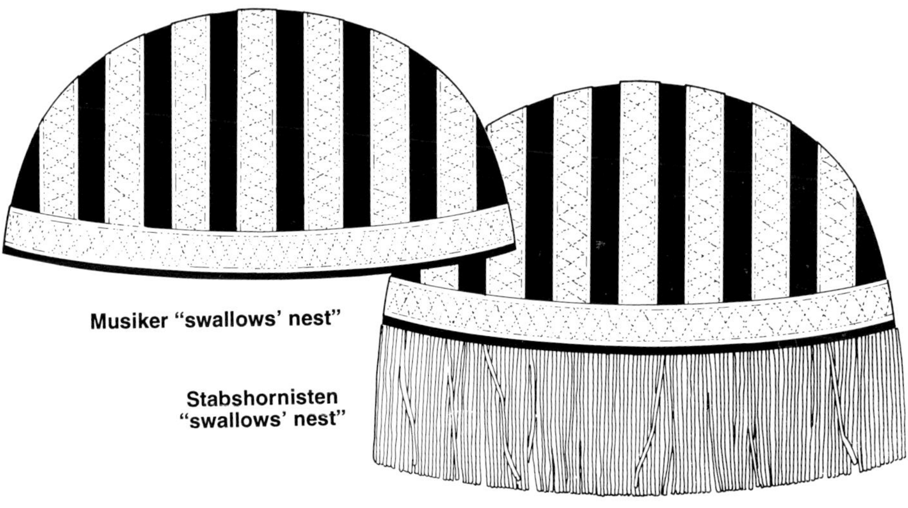

Musiker "swallows' nest"

Stabshornisten "swallows' nest"

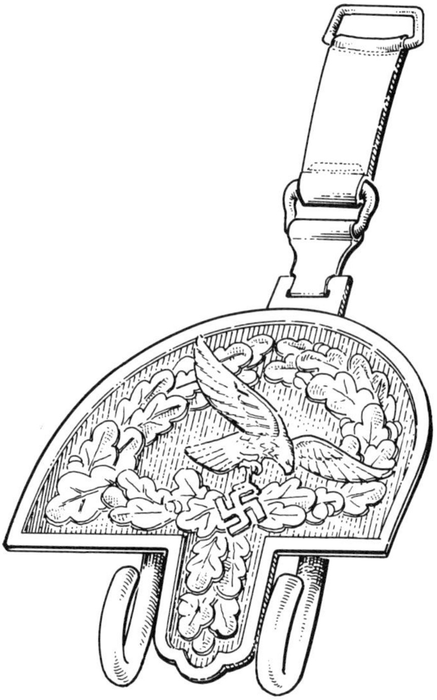

Snare drum hanger

Regiment "General Göring" Spielleute in a military parade in Potsdam (May 23, 1938), shortly before thei historical trip to Italy with the Führer.

Kettle Drum Banners

Luftwaffe kettle drum banners consisted of five panels, four of which were blue-grey and trimmed in aluminum braid. The fifth panel consisted of an aluminum brocade Luftwaffe eagle on a base of Waffenfarbe (red, yellow, or brown). The lower edge of the banner was adorned with an aluminum fringe and tassels (see photo at right).

Trumpet Banners

Trumpets utilized by Luftwaffe bands were decorated with distinctive banners, whose base color was in Waffenfarbe (red, yellow or brown). A Luftwaffe eagle in aluminum brocade was positioned on the right side of the banner. The details of these eagles were emphasized with black embroidery bordering along the feathers, the beak, claws and swastika. The banner's reverse or left side displayed the unit or garrison designation in aluminum or white embroidery. Its size was 50 x 39 cm.[11]

[11] Luftwaffen-Verordnungsblatt, March 16, 1936. The order was numbered 316 and dated March 6, 1936.

Trumpet banner (obverse)

(reverse)

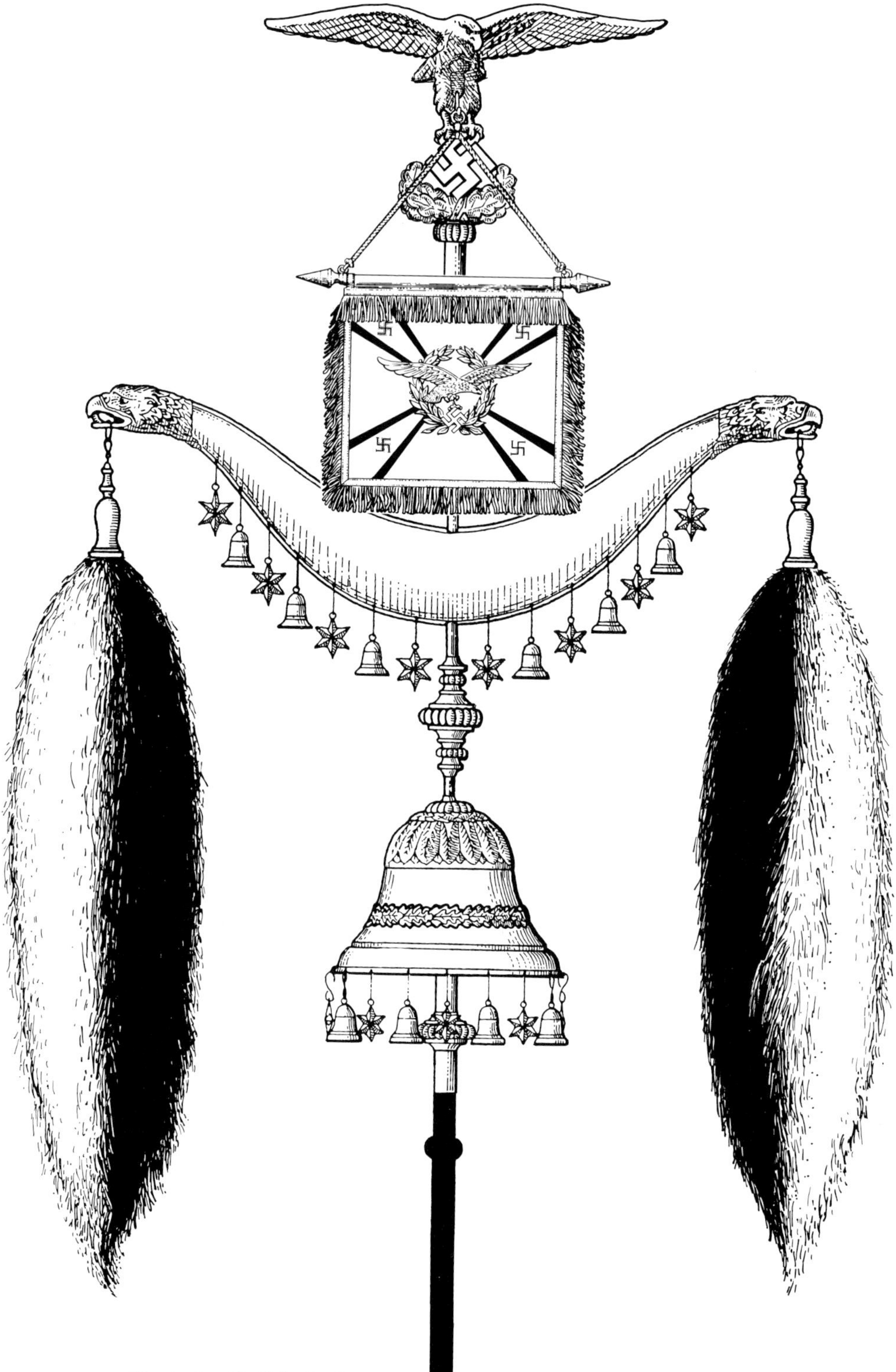

The Luftwaffe Schellenbaum

The Schellenbaum (Jingling Johnnie)

The Schellenbaum was one of the most ornate accessories utilized by German marching units and was carried at the head of most military bands. It was frequently presented to a unit by the inhabitants of the garrison town, who had purchased it privately. The small Schellenbaum flag was in the appropriate Waffenfarbe as that of its unit and was trimmed with aluminum embroidery, braid and fringe. The unit or garrison designation was on the reverse of the flag in either aluminum or white embroidery.

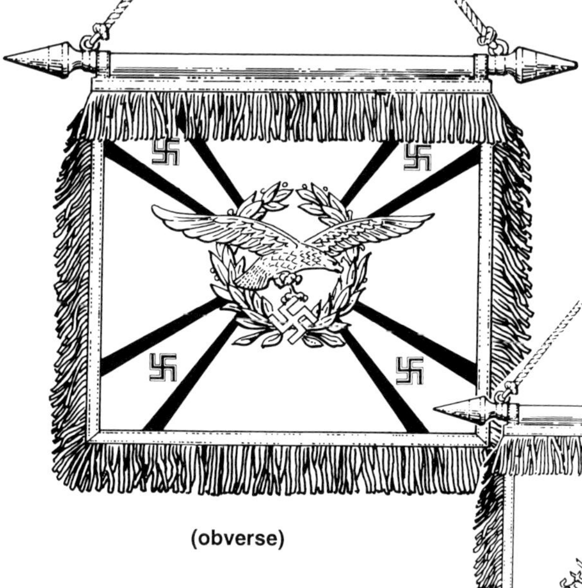

(obverse)

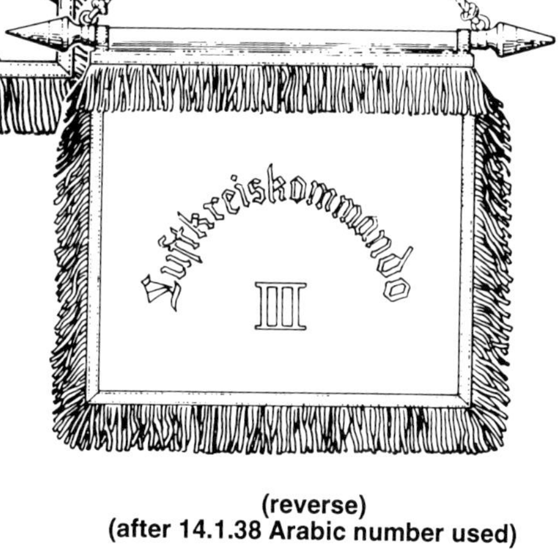

(reverse)
(after 14.1.38 Arabic number used)

Note the variation Schellenbaum for Regiment "General Göring."

Fire Fighting Branch of the Luftwaffe

The Luftwaffe's fire fighting branch assigned to air bases consisted of:

(a) Fire fighting officials of the Luftwaffe
(b) Civilian employees (salaried and hourly)
(c) Luftwaffe personnel (attached as firemen)

The Officials' Uniform

High ranking officials of the Luftwaffe's Feuerschutzbeamte wore the uniform of the Wehrmachtbeamte der Luftwaffe (Officials of the Luftwaffe). Fire fighting officials of lower grades (i.e. Hauptbrandmeister, Oberbrandmeister, Brandmeister and Unterbrandmeister) wore a uniform closely resembling the Feuerwehr uniform.[12] The peaked cap was dark blue with a black cap band and black peak. Above and below the cap band and around the crown of the cap was a 2 mm wide piping in crimson. An aluminum hand-embroidered Luftwaffe eagle and winged rosette, plus an aluminum colored chin strap, was worn by officials.

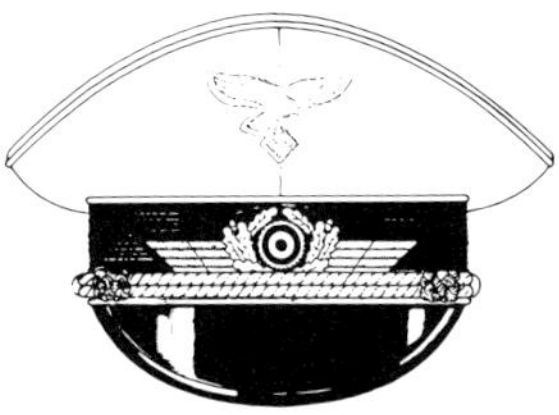

Officials' peaked cap

The closed-collar tunic was dark blue (as for Feuerwehr), with one row of eight aluminum pebbled buttons and an aluminum hand-embroidered Luftwaffe eagle over the right breast pocket.[13] The rank insignia (shoulder straps and collar patches) were as for Luftwaffe officials with the exception of the base color being dark green. The collar patches for officials of lower grades were piped in a 2 mm wide alternating aluminum and blue-grey cord. With this tunic were worn long black trousers with crimson piping down each leg.

The issue great-coat was in black with a dark-blue collar piped in crimson, and similar in cut to the naval officers' model with two rows of six aluminum buttons. The dark green and golden-brown based shoulder straps were worn on the great-coat. A cape was also issued in black with a dark blue collar piped in crimson.

[12] Hettler, *Uniformen der Deutschen Wehrmacht (Nachtrag 1939/1940)*, p. XLV. As of late 1939, all fire fighting officials were to wear the standard Luftwaffe officials' uniform (Wehrmachtbeamtenuniform der Luftwaffe). The dark-blue Feuerwehr-style uniform continued to be worn by the civilian employees.

[13] Luftwaffen-Verordnungsblatt, April 5, 1937, #407. The lowest officials rank of Unterbrandmeister had the Luftwaffe breast eagle in matte-grey thread whereas all higher officials had the aluminum hand-embroidered version.

Official's uniform in the rank of Unterbrandmeister

Unterbrandmeister

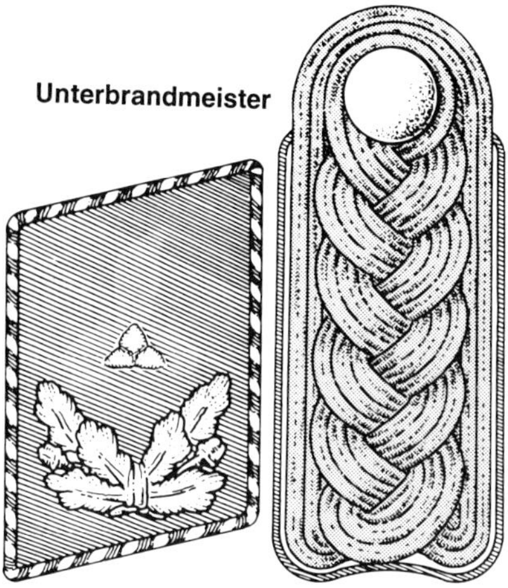

Before November 2, 1940:

Unterbrandmeister - as illustrated (one matte-grey triangular pip on the collar patches) and with black chin strap. **Brandmeister** - as illustrated but with two aluminum triangular pips on the collar patches and one aluminum pip on the shoulder straps and aluminum cap cords. **Oberbrandmeister** - as illustrated but with three aluminum triangular pips on the collar patches and two aluminum pips on the shoulder straps and aluminum cap cords. **Hauptbrandmeister** - as illustrated but with four aluminum triangular pips on the collar patches and two aluminum pips on the shoulder straps and aluminum cap cords. The outer blue-grey cord of this rank's shoulder straps had entwined aluminum chevrons.

Unterbrandmeister and other officials (leather chin strap)

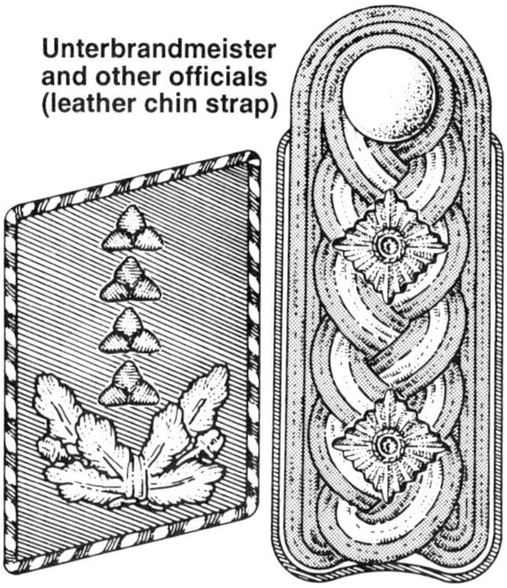

After November 2, 1940:

From this date, fire fighting officials wore the standard Wehrmachtbeamten uniform with their second branch color being red (no longer golden-brown).

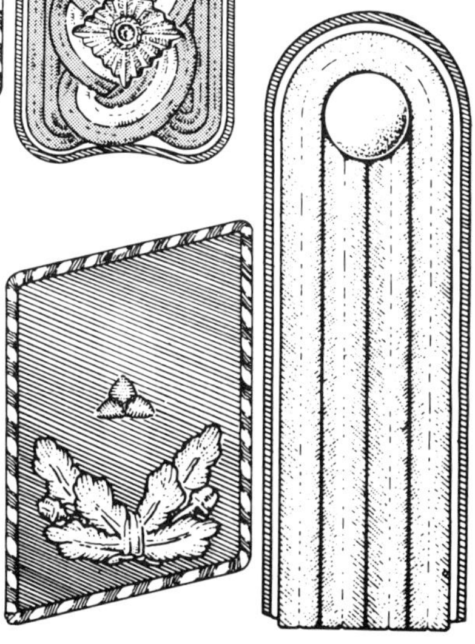

Brandmeister and other officials (aluminum cap cords)

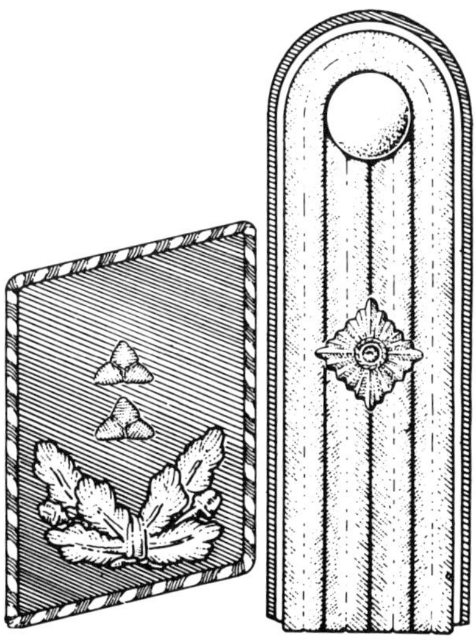

Oberbrandmeister, Hauptbrandmeister, and other officials (aluminum cap cords)

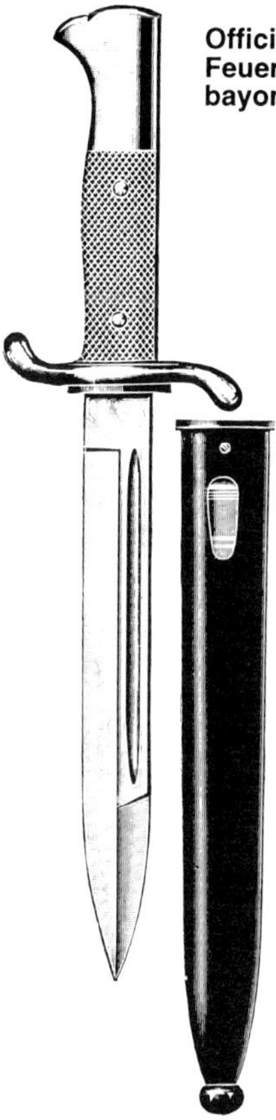

Officials' Feuerwehr-style bayonet

While on duty, officials were permitted to wear a bayonet suspended from a dark-brown belt for Unterbrandmeister and from a light-brown belt with cross-over strap for Brandmeister and above. Off-duty officials and civilian personnel wore a special sabre corresponding to that of the Army pattern.

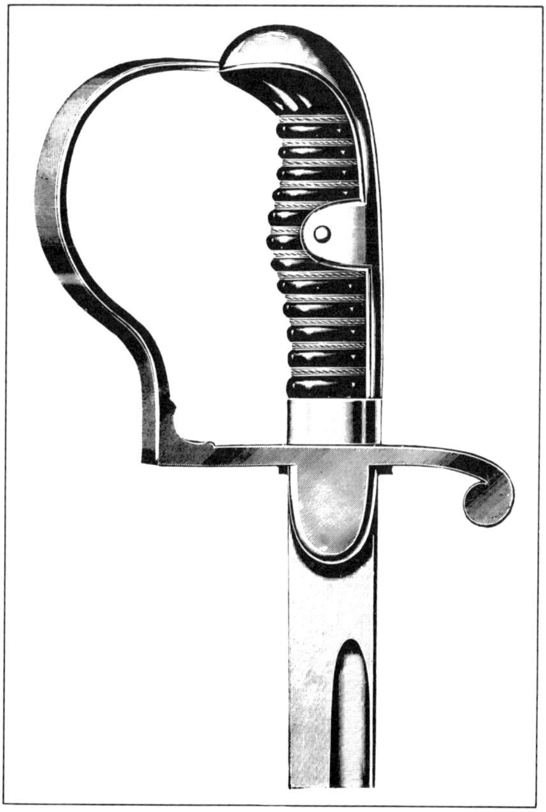

Sabre worn by ranks of Hauptbrandmeister and below.

The Civilian Employees' Uniform

The civilian employees' uniform was identical in cut and color to that of the official's above. An outstretched national eagle was positioned on the upper left sleeve, however, instead of a Luftwaffe eagle above the right breast pocket. The collar patches were black, piped in crimson, and with aluminum insignia. The shoulder straps were dark blue and also piped in crimson.

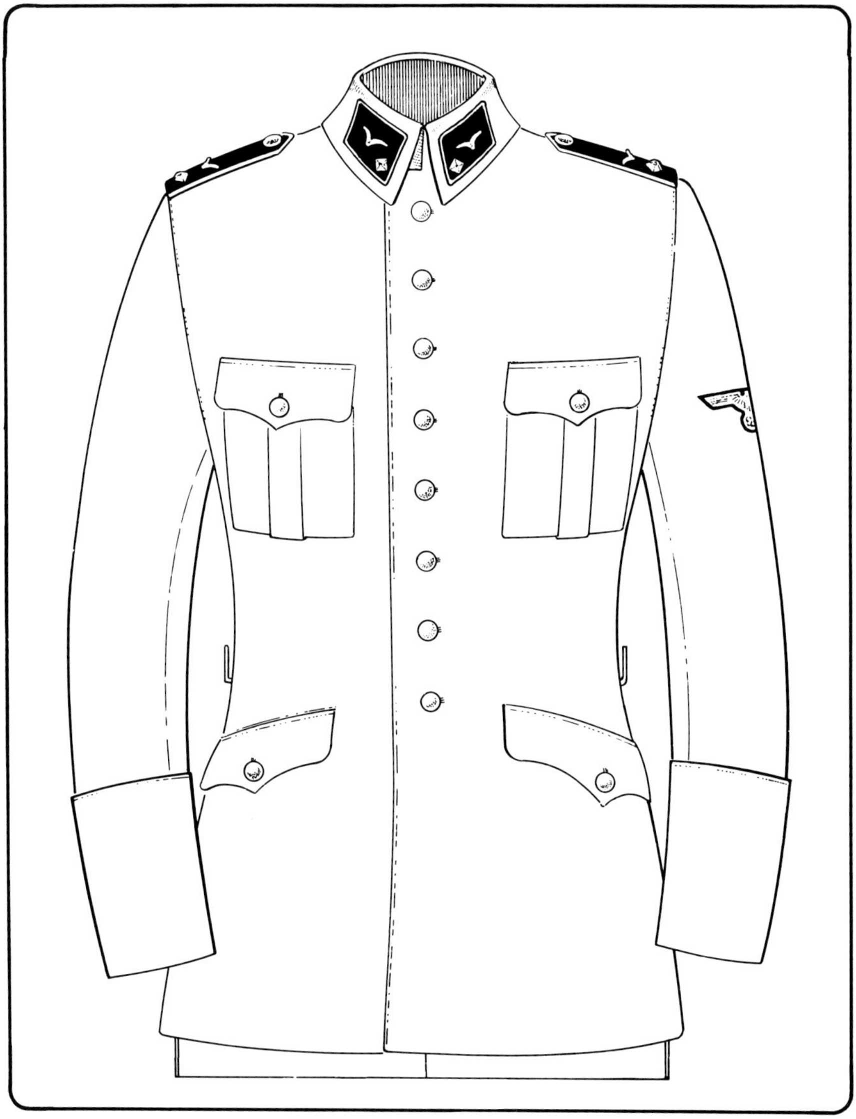

Civilian employees' uniform in rank of Oberfeuerwehrmann

Feuerwehrmann

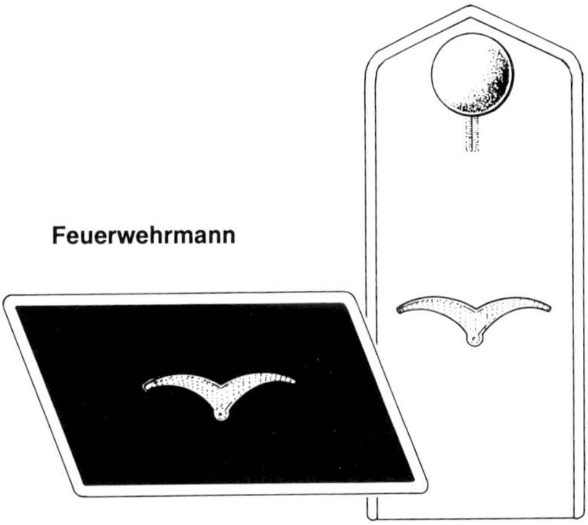

Oberfeuerwehrmann

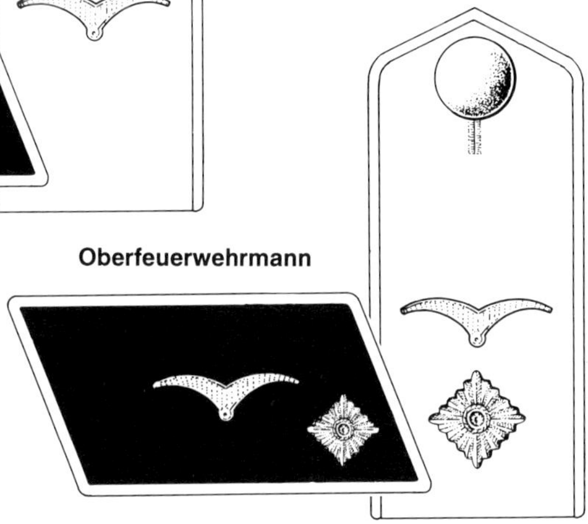

Löschmeister

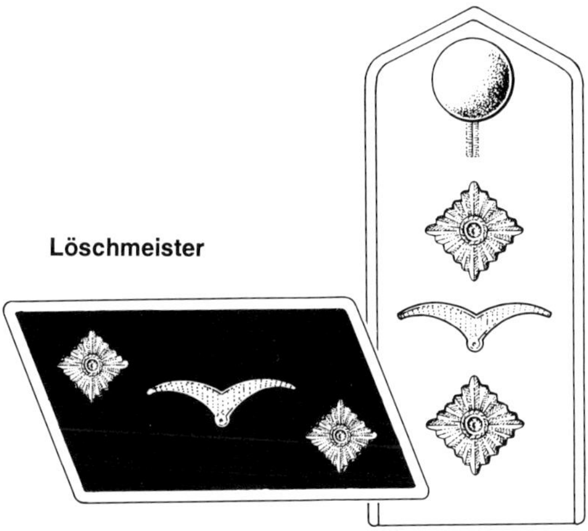

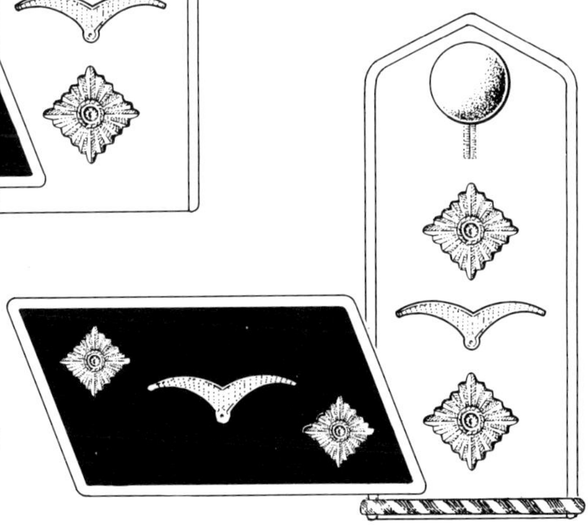

Löschmeister with aluminum and blue-grey cord on end of strap indicating a candidate for an officials' grade.

Civilian Employees' Peaked Cap

The peaked cap was also identical in color to that of the officials, but with black leather chin straps. Its insignia, however, was different with a national eagle positioned on the front of the crown and a winged civilian employee insignia on the cap band (both in aluminum).[14]

The civilian employee, going to and from work, was permitted to wear a black belt with a distinctive belt buckle and cross-over strap. A short firemen's bayonet with knot was also worn. While on duty, however, no sidearms were permitted to be worn.

A one-piece work suit in black, which was unlined, without insignia and similar to that of the Luftwaffe's, was also issued. The work cap was also in black with a matte-grey, machine-embroidered national eagle on front, and again similar in cut to that of the Luftwaffe's forage cap. During active fire fighting, a firemen's belt with large hooks, axe, long rope and other fire fighting equipment was worn.

A light-weight, black helmet, similar to that of other firefighting units, both professional and volunteer, was also issued. The national colors (black/white/red) were in a shield on the right side of the helmet and an aluminum colored Luftwaffe eagle on the left.

Uniform of Luftwaffe Personnel attached as Firemen

These attached Luftwaffe personnel wore their own Luftwaffe uniform but with a 13 cm wide, light blue armband on each upper sleeve. Each armband was marked with the appropriate garrison headquarters stamp.

[14] Luftwaffen-Verordnungsblatt, February 7, 1938, #34. This specially designed emblem was to be worn (besides the civilian personnel in a fire fighting capacity) by civilians employed as doormen, elevator operators, guards, drivers and gatemen at specific Luftwaffe posts.

Police-style buckle worn by civilian employees

NOTE:
The helmet illustrated below also exists with an aluminum comb. Black Model 35 helmets were also issued with Luftwaffe insignia and with or without combs.

Feuerwehr-style helmet

Luftwaffe Forestry Service

Luftwaffe Forestry Officials were under the supervision of the State Forestry Service, whose task was the regulated conservation of Germany's timberland and wildlife. They were, however, solely responsible for Forestry Service within the Luftwaffe. They wore the uniform of the State Forestry Service (but with black Waffenfarbe instead of dark green) and were classified by Forestry designations instead of military ranks.

The Forestry Uniform

The peaked cap was in grey-green with the capband and piping in dark green. The Luftwaffe eagle and winged Reich cockade (embroidered) were positioned as on the standard peaked cap. Cap cords for Unterförster and below were of black patent leather, for higher officials in aluminum and for Oberlandforstmeister in gold.

The forage cap was grey-green with wool embroidery insignia. A cap in the cut of the mountain cap was also available, in grey-green with the Luftwaffe eagle over the Reich cockade (wool embroidery). A felt Forestry officials' hat, including its hatband and border, were all in green. The Luftwaffe eagle was positioned above the hatband and was in the same color embroidery as on the wearer's collar patches.

Felt Forestry officials' hat

The Forestry officials' tunic was in two styles, which were as follows:

(a) a grey-green tunic with dark-green collar and green buttons for forestry service and garrison duty, and...

(b) a tunic with dark-green velvet collar and aluminum colored buttons (gold colored buttons for Oberlandforstmeister) for reporting dress, parade dress and association dress (for social affairs).

The cut of the tunic was as that of the Luftwaffe's uniform tunic, except there were two buttons above the vent in back. The Luftwaffe eagle was positioned above the right breast pocket in grey wool embroidery up to Unterförster, aluminum for higher officials and in gold for Oberlandforstmeister. The breast eagle was on a dark green backing. During the summer, the tunic (a) was available in either linen or ticking.

A forest blouse was also issued which was in the same cut as the Luftwaffe flying blouse (with the undress uniform). The color and insignia was as (a) above. A white tunic was worn by officials from Förster and up, and in the same cut as that of the Luftwaffe's white tunic. The insignia was as in (a). An elegant evening dress uniform, in grey-green with a dark green velvet collar and in the cut of the Luftwaffe's evening full-dress uniform, was worn on special occasions. Oakleaf embroidery (aluminum) decorated the dark green velvet collar. Pant stripes were as for Luftwaffe officers.

A dark green shirt and tie were worn with the service dress, and a light green shirt and dark green tie were worn with the garrison duty dress. A white shirt and black tie were worn with the reporting and parade dress (also with the summer dress). A white frock-shirt and bow tie were worn with the frock-coat.

The great-coat was in grey-green with dark green buttons and in the Luftwaffe cut. The collar was in dark green (the lapels for Oberlandforstmeister were also in dark green). A grey-green cape with dark green collar was also worn. The Luftwaffe eagle, in aluminum, was worn with it (in gold for Oberlandforstmeister).

The straight pants were in grey-green with dark green piping down each leg (the Oberlandforstmeister rank had double pants stripes). Ski pants, short pants and breeches were also in grey-green but without the green piping. Knee-length shorts, in mouse-grey cloth or leather, were worn with the undress uniform while at mountain hunting preserves.

The aiguilette was worn with the parade dress, full-evening dress and on the frock-coat by ranks of Förster and up.

Forstaufseher

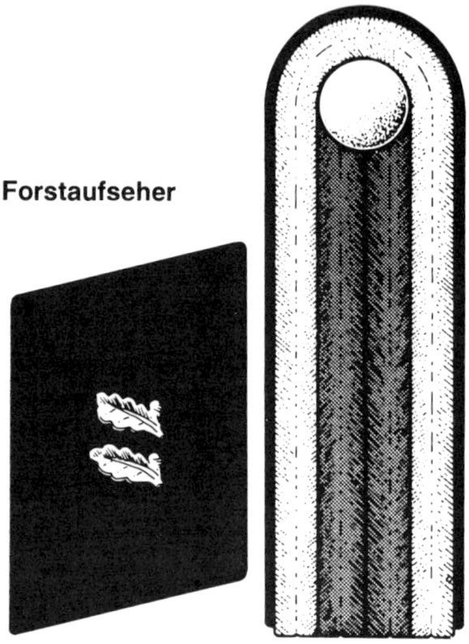

Hilfsförster

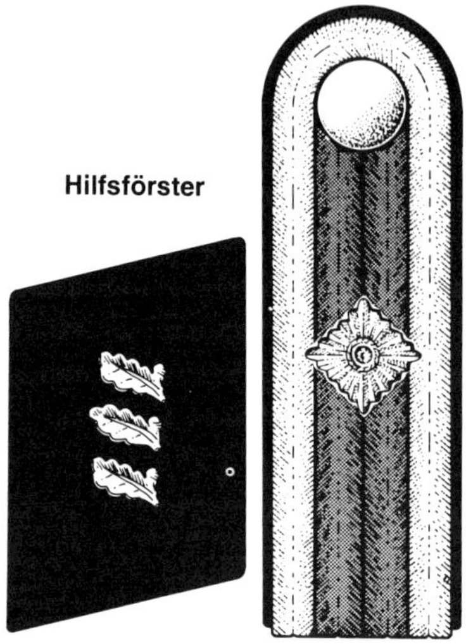

Unterförster

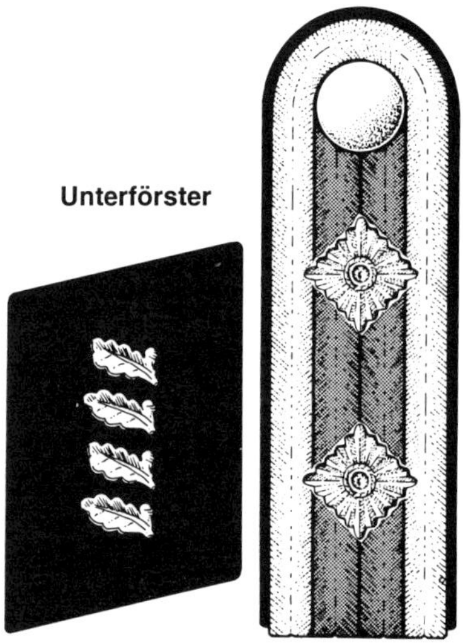

The three service grades above had aluminum stamped-metal insignia on their peaked caps, metal oakleaves on their collar patches and shoulder straps with the outer flat braid in an aluminum-green and the inner flat braid in dark green. Pips for the shoulder straps were in a gold-colored metal.

Stag-grip Forestry knife for Unterförster

Förster

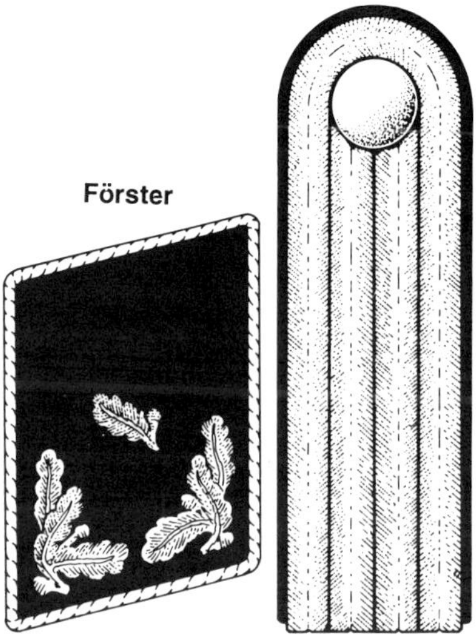

Revierförster

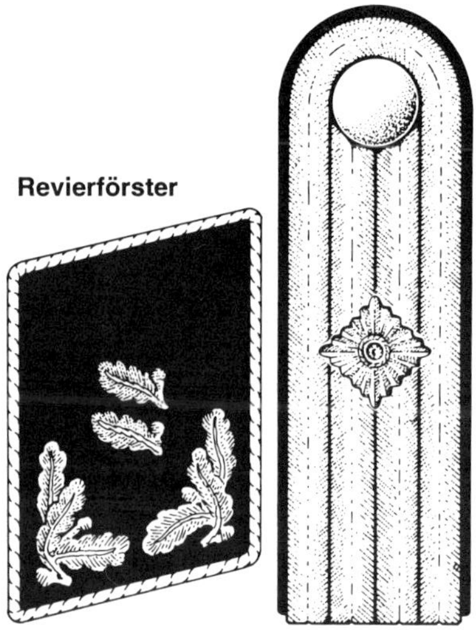

Oberförster

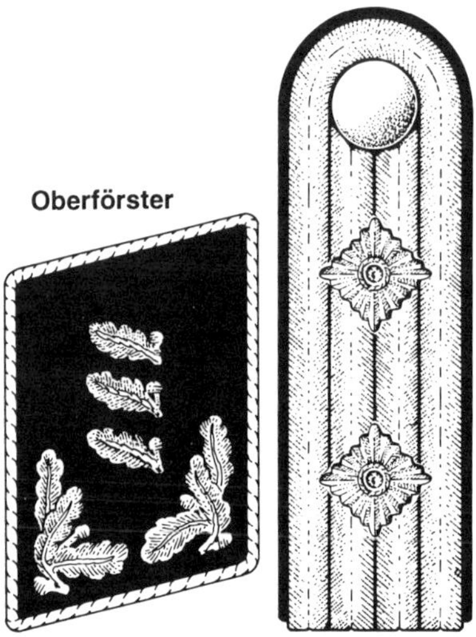

Forstamtmann

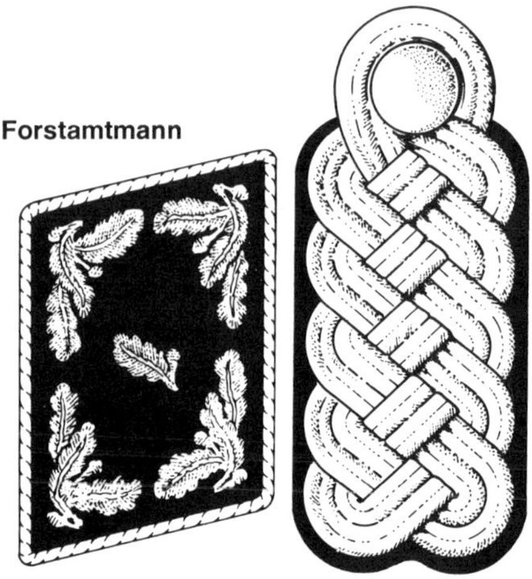

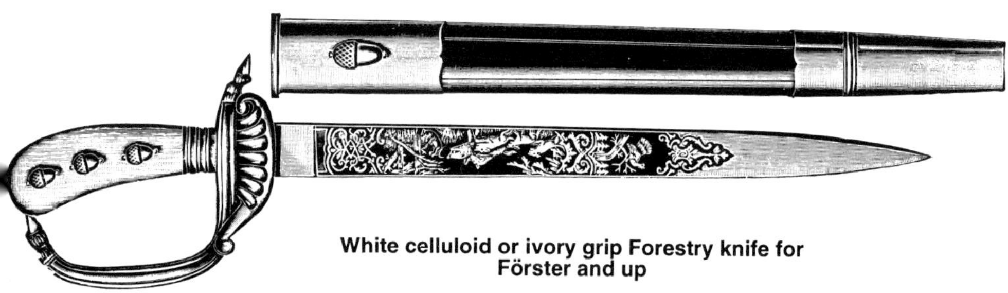

White celluloid or ivory grip Forestry knife for Förster and up

Forstassessor

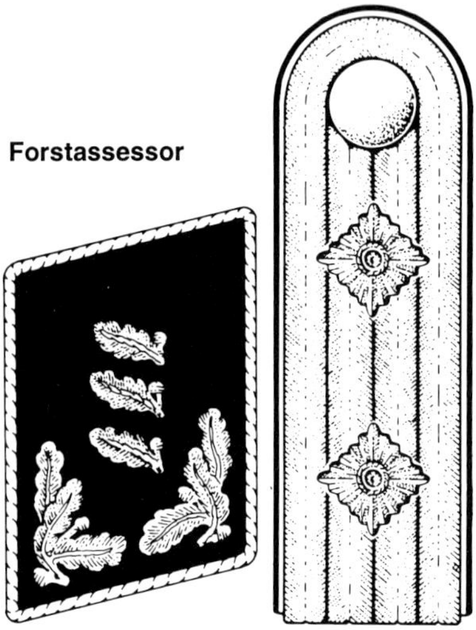

Forstmeister

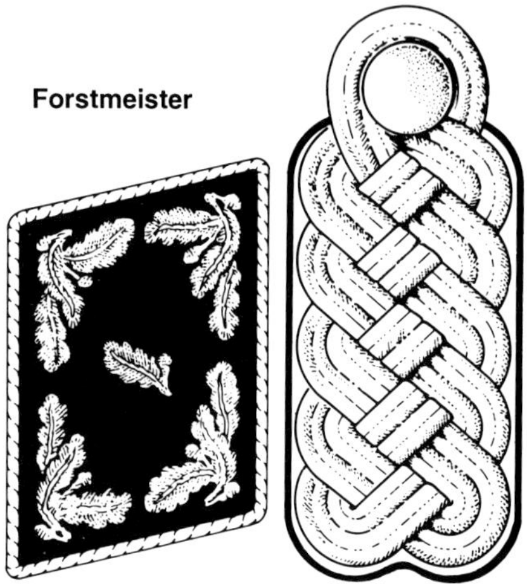

Oberforstmeister

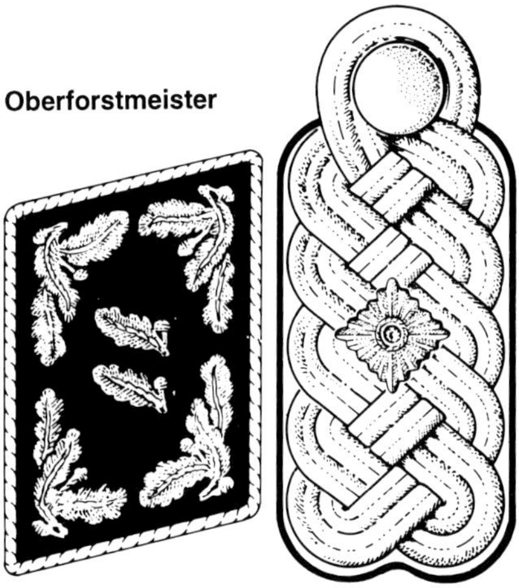

Landforstmeister

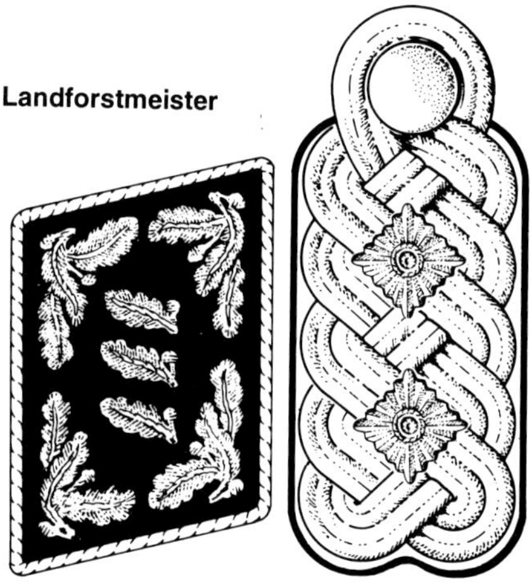

These service grades had aluminum-embroidered insignia on their peaked caps (on dark green base), shoulder straps with aluminum-green braid and gold-colored pips, 2mm aluminum collar patch piping and aluminum-embroidered oakleaves, and aluminum collar piping for the tunic (a) and (b), evening dress frock-coat, and great-coat.

NOTES:

1. An inner border in bright-aluminum was worn on the shoulder straps of higher officials (Forstassessor-Landforstmeister).[15]

2. collar patches for the tunic (a) and the great-coat were in black cloth and for tunic (b) in black velvet.

3. Shoulder straps for the tunic (a), the forest blouse, and great-coat were matt and for the tunic (b) and the evening frock-coat in a bright finish.

The shoulder straps for Oberlandforstmeister were as for a general's rank (two gold and one aluminum-green cord). All other insignia was in gold.

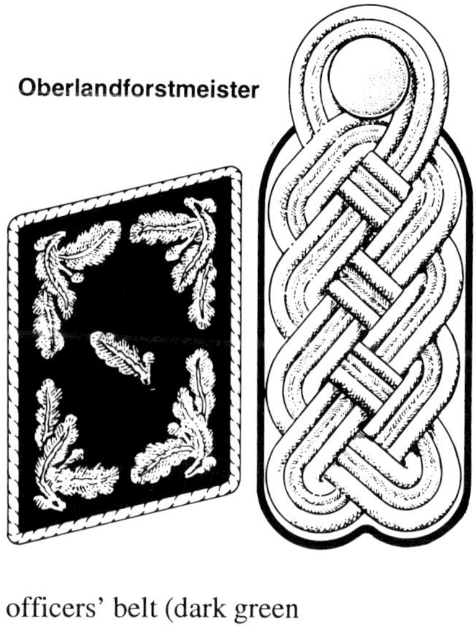

4. Officials wore the Luftwaffe officers' belt (dark green leather) with cross-over strap (Förster and up). The wearing of the cross-over strap was, however, discontinued in early 1940. In mid-1941, light brown leather belts and holsters were then to be worn in place of dark brown.[16]

The State Forestry knife was worn, with stag-grip for low ranking officials and with white celluloid or ivory grips for senior officials. Three grades of knots distinguished the rank of the Forester, which were as follows:

(a) Unterförster and below - dark green knot with dark green tassel
(b) Förster and up - the aluminum portepee
(c) Oberlandförster - also the aluminum portepee

The Luftwaffe officers' dagger could also be worn in place of the Forestry knife by ranks of Förster and up, plus the aluminum portepee.

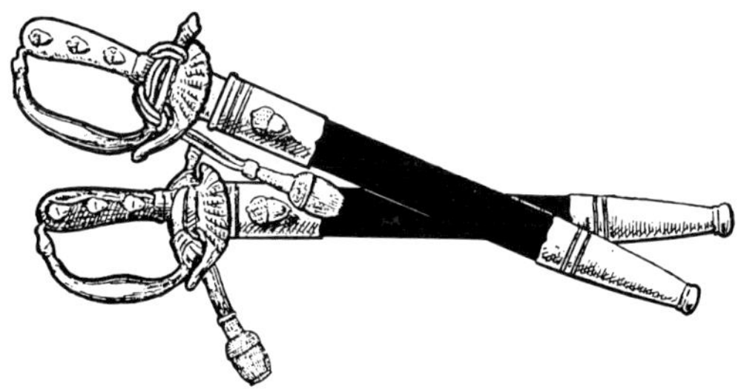

[15] Hettler, E., *Uniformen der Deutschen Wehrmacht, Nachtrag 1939/1940.* Excellent coverage is to be found on the Luftwaffe Forestry Officials on pages XLVII - XLIX.

[16] Luftwaffen-Verordnungsblatt, July 21, 1941. The order was numbered 791 and dated July 10, 1941.

Generalluftzeugmeister Personnel

Generalluftzeugmeister personnel were civilian technicians, whose specialties were usually in maintenance or supply. The Generalluftzeugmeister was authorized to send them to occupied countries on special assignments where their skills were required. Those civilians already stationed outside the Reich were also under his supervision.

Special insignia was initiated for these civilian personnel who were, however, governed by military regulations. The Luftwaffe eagle on a cogged wheel came in two sizes, which were as follows:

size 1 - 35mm (wingspan dimension), for wear on the lapel of the civilian jacket

size 2 - 65mm (wingspan dimension), for wear on the peaked cap of a technical officer's uniform

Black on golden-yellow GL insignia for wear on the right breast of coveralls. A black on yellow "General-Luftzeugmeister" armband was also worn for further identification on civilian dress.

TN Personnel

(Assigned to the Luftwaffe)

In 1942, TN-(Technical Emergency Corps) field units were organized to operate in the German rear areas. TN members assigned to the Luftwaffe used their professional and technical skills to establish, maintain and repair vital defense installations. The standard

The TN leader (above and below) is wearing the Luftwaffe forage cap, flying blouse, pants and belt with buckle. All other insignia is TN.

TN cuff title

Luftwaffe uniform was worn with the addition of the TN rank insignia, arm eagle and cuff title.

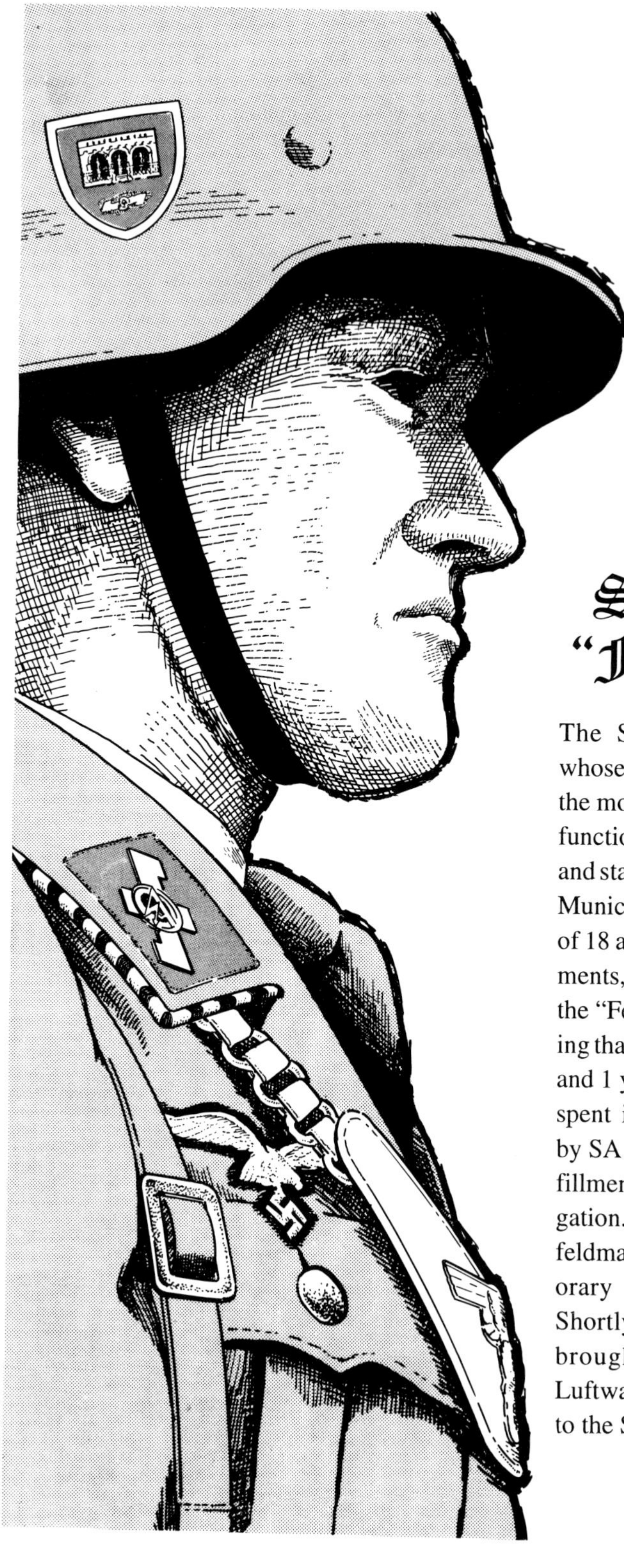

SA-Standarte "Feldherrnhalle"

The SA-Standarte "Feldherrnhalle" whose headquarters was in Berlin, wa the most elite organization of the SA. It function was to guard SA garrisons, Part and state offices and the Feldherrnhalle i Munich. Party members between the age of 18 and 25, who had passed set require ments, could enlist for a 3-year period i the "Feldherrnhalle." The stipulation be ing that they had served 1/2 year in the SA and 1 year in the Hitler Youth. This tim spent in the Standarte "Feldherrnhalle" by SA personnel was considered the ful fillment of their peacetime military obli gation. On January 12, 1937, General feldmarschall Göring was appointed hon orary Chief of the "Feldherrnhalle." Shortly thereafter, the Standarte wa brought under the auspices of the Luftwaffe, with direct responsibility onl to the SA-Stabchef.

During the Sudeten campaign in October 1938, the Standarte formed the Luftlande-Regiment "SA-Standarte Feldherrnhalle" (airborne regiment) within the 7. Flieger-Division of the Luftwaffe. From this time up to the spring of 1939, Luftwaffe rank designations were utilized.

When the war broke out in 1939, the many Sturmbanne (battalions) which made up Standarte "Feldherrnhalle" began a strenuous training program designed to make them specialists in varied military fields. Several of these SA-Sturmbanne, such as those at Standorte (garrisons) Berlin, Munich, Krefeld, Stettin, Stuttgart, Hattingen and Vienna, concerned themselves exclusively with Luftwaffe training. They were completely equipped by the Luftwaffe and received their field training from Luftwaffe instructors. From these Luftwaffe oriented groups came the first SA volunteers (forming Fallschirmjäger-Regiment 2) who fought in the ensuing battles of Eben Emael, Rotterdam and Crete.*

Later in the war, the bulk of Standarte "Feldherrnhalle" was drawn into the Luftwaffe or the Heer, while the remainder was entrusted with the task of teaching marksmanship and other specific military duties to SA men who were going to join the Army, Navy, Waffen-SS or Luftwaffe.

*SA-Standarte "Feldherrnhalle" personnel also served in the 60.Pz.Gren.Div., Minensuchflottille "Brill" and the 18.SS-Pz.Gren.Div. "Horst Wessel."

(Above) "Feldherrnhalle" gorget worn by all members during the pre-war years. (Right) "Feldherrnhalle" metal insignia for the collar patch.

The standard SA service uniform (brown) was worn for parades and on-duty hours, with the special "Feldherrnhalle" collar insignia (crimson Waffenfarbe), crimson and white collar piping, two SA shoulder straps (crimson underlay during the peacetime years and brown with silver chevrons on a grey underlay during the war, for Unterführer and men), standard Party armband, aluminum on medium brown "Feldherrnhalle" cuff title, and the Luftwaffe breast eagle. For field service and military maneuvers, the standard blue-grey Luftwaffe uniform

This SA-Sturmführer, from Gruppe Kurpfalz (wearing the SA service tunic) saw active duty with the Luftwaffe and was a holder of the German cross in gold.

"Feldherrnhalle" cuff title

was worn with a "Feldharrnhalle" cuff title (aluminum on medium brown). The helmet which was worn (brown with the SA uniform and blue-grey with the Luftwaffe uniform), had a Luftwaffe eagle on the left side and the "Feldherrnhalle" monument within a shield on the right (silver and crimson).

Right helmet decal

Left helmet decal

SA-Sturmmann to SA-Haupttruppführer (white metal runic insignia, gold for SA-Sturmführer and up).

Runic insignia worn on both shoulder straps by "Feldherrnhalle" personnel in armed forces (circa May 1944).

NSKK/Luftwaffe Transport Units

Early in the war, NSKK units were attached to the Luftwaffe and the Army in a transportation role.[1] As the war progressed, other NSKK transport units were formed...among these were "NSKK-Transportgruppe Todt"* and "NSKK-Transportbrigade Speer" ("Todt" wore olive-green uniforms with NSKK insignia and "Speer" wore blue-grey uniforms with NSKK insignia). Their composition and official titles varied from time to time, until late 1944, when they were all absorbed under the overall title of "Transportkorps Speer" (Transport-Regimenter 2 and 3 were attached to the Luftwaffe and the remainder of the Korps to the Army).

At this time, the uniform for those units which were tactical detachments of the Luftwaffe was in blue-grey. The NSKK eagle was worn on the upper left arm and the Luftwaffe eagle on the right breast (on an olive base). Standard NSKK rank insignia was worn with this uniform (collar patches (black) and shoulder straps on an olive base).

*"NSKK-Transportgruppe Todt" had no association whatsoever with "Organization Todt."

The "Transportkorps Speer" Uniform

Late in 1944, a new "SP" uniform was introduced and the continued wearing of Army, Luftwaffe and NSKK uniforms in the recently reorganized transport organization was forbidden. The uniform was of the same cut for all ranks and was in an olive color.

Leaders wore a field blouse in olive colored Trikot, which had six aluminum or olive-colored buttons, two breast and two side pockets and a dark brown collar. The collar patches and shoulder straps were on a colored underlay in the appropriate Waffenfarbe. The Luftwaffe eagle, on an olive colored base, was positioned above the right breast pocket.

The riding breeches were also in olive for ranks up to Stabskapitän. Leader ranks from Brigadekommandant to Führer im Korpsstab wore two 4 cm wide stripes plus a 0.2 cm center piping on each leg of the breeches. Bloused pants were worn by leaders only in Italy or the northern regions. Straight pants were also permitted to be worn by leaders during transport duty, for Walking-Out Dress and when on base. The two 4 cm wide stripes and 0.2 cm center piping (dark brown) were also worn on the straight pants by ranks from Brigadekommandant to Führer im Korpsstab.

[1] One such unit was "NSKK Regiment Luftwaffe," which not only supplied bombs to several airfields in France (1940-41) but also assisted in the construction of airfields for the battle against England. The blue-grey uniform was worn, but with NSKK insignia.

The leaders' great-coat was in olive with a dark turn-down collar without collar patches. Shoulder straps were worn with this double-breasted coat. Korpskommandant, Brigadekommandant and Gruppenkommandant ranks wore the top three buttons of their great-coat open, with the exposed lapels in dark brown. An olive colored leather coat and rain coat could also be worn.

The leaders' peaked cap was in olive with a dark brown cap band, metal or embroidered Luftwaffe eagle and the winged Reich cockade. Aluminum cap cords were worn. Ranks of Brigadekommandant and above wore the Luftwaffe eagle, winged Reich cockade and cap cords in gold. Leaders were also permitted to wear a mountain-style cap in black with an embroidered Luftwaffe eagle and Reich cockade. Silver piping was worn around the top of the cap by leaders while ranks of Brigadekommandant and above wore gold piping.

A black leather belt with double-claw buckle was worn as well as a black leather holster on the left hip. Gloves were either in grey or brown leather.

The "SP" collar patches for leaders were on a base of appropriate Waffenfarbe with the "SP" in aluminum colored thread and piped with aluminum cord. Ranks of Brigadekommandant and above had their collar patch embroidery and piping in gold on a Waffenfarbe underlay.

NOTE:

An olive colored shirt could be worn in place of the tunic during the hot season. It was available in long or short sleeves and worn with a tie of the same color.

The uniform (tunic, pants and great-coat) for NCO equivalents (Unterführer) and men was identical in cut and color to that for leaders, but with appropriate insignia. The mountain-style cap for these ranks was in black, with the Luftwaffe eagle and Reich cockade on a black underlay.

A special traveling suit, in light, olive colored ticking material, was issued for wear during the hot season of the year. The first version was a one-piece coverall with shoulder straps and arm rank insignia (no collar patches). As an added identification to the light weight traveling suit, metal block letters ("SP") were positioned on the shoulder straps. The Luftwaffe eagle was positioned over the right breast. The suit was also available in a two piece version with two large breast pockets on a short jacket and baggy pants which could be worn over the uniform pants.

The belt worn was in black leather with a one or two prong buckle. Black pistol holsters were also worn, but only by NCOs in the position of Zugführer.

Additional identifying insignia worn by men of "Transportkorps Speer":

(a) Transport sergeants (Schirrmeister) wore a circular specialty patch with an embroidered "S" on a dark brown base (lower left sleeve).

(b) Medical personnel wore a circular specialty patch with an embroidered AEsculapius Staff on a dark brown base (lower left sleeve).

(c) The duty NCO wore the golden-yellow lanyard over the right shoulder (see p.211).

(d) Instructors wore from the second button of their service tunic to the right breast pocket, a signal whistle and cord (brown).

Waffenfarben

NCO equivalents and men - black
Troop Supply - light blue
Technical - pink
Judicial Service - wine-red
Medical service - dark blue
Administrative Service - green
Sonderführer - pale violet-grey

Rank Insignia

The collar patches for NCO equivalents and men were identical in design. They had the "SP" in mouse-grey, woven block letters on a black underlay. Additionally, the following were worn on the upper left arm of the field blouse and great-coat:

(a) Oberkraftfahrer rank - 0.9 cm aluminum chevron

(b) Hauptkraftfahrer rank - two 0.9 cm aluminum chevrons

Whereas the Kraftfahrer rank had plain olive colored shoulder straps with a 2 mm wide piping in appropriate Waffenfarbe, Unterführer grades had in addition a 0.9 cm aluminum Tresse around the shoulder straps (see illustration) and a 1.5 cm aluminum Tresse around the collar of their field blouse.

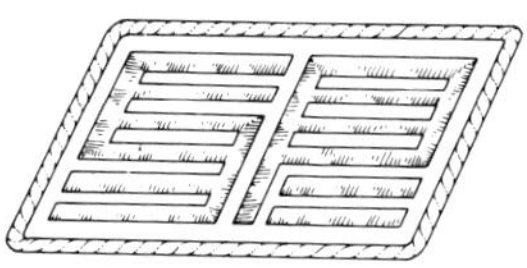

The collar patches for leaders had the "SP" in hand-embroidered aluminum colored thread and aluminum piping. The ranks of Brigadekommandant and up wore the same collar patches but with gold embroidery and gold piping.

1 2 3 4 5 6 7 8

High leaders:

1. Korpskommandant
2. Gruppenkommandant
3. Brigadekommandant, Generalingenieur, Generalintendant, Generalarzt, Generalrichter

Leaders:

4. Oberstkapitän, Oberstingenieur, Oberstintendant, Oberstarzt, Oberstrichter
5. Oberfeldkapitän, Oberfeldingenieur, Oberfeldintendant, Oberfeldarzt, Oberfeldrichter
6. Oberstabskapitän, Oberstabsingenieur, Oberstabsintendant, Oberstabsarzt, Oberstabsrichter
7. Stabskapitän, Stabsingenieur, Stabsintendant, Stabsarzt, Stabsrichter
8. Oberfeldkornett, Oberingenieur, Oberzahlmeister, Oberarzt
9. Feldkornett, Ingenieur, Zahlmeister, Assistenzarzt

NCO equivalents (Unterführer) and Men:

10. Stabsfahrmeister
11. Hauptfahrmeister (also double aluminum rings around both sleeves)
12. Oberfahrmeister
13. Fahrmeister
14. Unterfahrmeister
15. Kraftfahrer

Exceptions

Administrative Leaders wore:

Verwaltungssekretär - as #9 but with green underlay and interwoven green and aluminum braid

Verwaltungsinspektor - as above but with one gold colored pip

Verwaltungsoberinspektor - as above but with two gold colored pips

Regierungsrat - as #6 but with green underlay and interwoven green and aluminum cords

Oberregierungsrat - as above but with one gold colored pip

Ministerialrat - as above but with two gold colored pips

Ministerialdirigent - as #3 but with green underlay and interwoven green, aluminum and gold cords

Sonderführer wore:

Sonderführer (Z)* - four aluminum cords positioned next to each other on a pale violet-grey underlay

NOTE:
Shoulder strap pips for High Leaders were in white metal, for Leaders in gold and for other grades in white metal.

9 10 11 12 13 14 15

Sonderführer (Z) in position of Oberfeldkornett - as above but with one gold cord slide on the base of the cords

Sonderführer (K)** - as above but with two gold cord slides

Sonderführer (B)*** - four small entwined aluminum cords on a pale violet-grey underlay

Leader grades carried regimental numbers plus specialist metal insignia on their shoulder straps:

(a) Administrative Supply Leader - metal administrative specialists' insignia
(b) Judicial Leader - metal judicial sword
(c) Technical Leader - metal "J"
(d) Medical Leader - metal AEsculapius Staff
(e) Administrative Leader - metal "V"

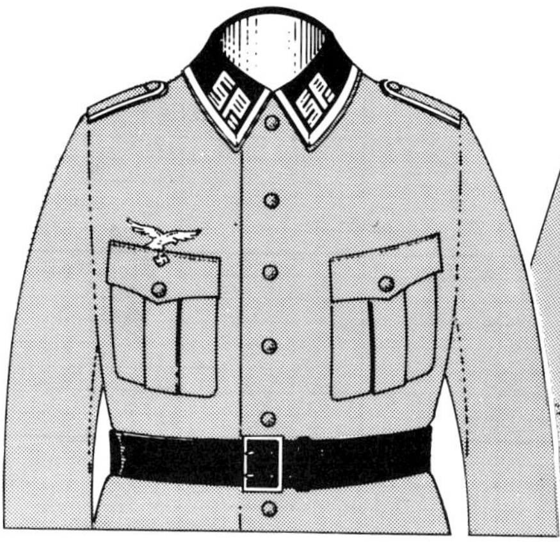

Service tunic (Unterfahrmeister)

Traveling suit (Hauptkraftfahrer)

* - appointment as Zugführer

** - appointment as Kompanieführer

*** - appointment as Bataillons - or Abteilungsführer

NOTES:

1. The uniform worn by foreign volunteers in "Transportkorps Speer" was identical to those worn by German nationals. The exception was an 8 cm national shield in the colors of the wearer's nation, worn on the upper left arm. Those without a definite nationality wore a white shield with no inscription.
2. The Führeranwärter (leader candidate) wore leader grade collar patches, silver piping around the top of the mountain-style cap and silver cap cords on the peaked cap. He also wore the Oberfahrmeister grade shoulder straps, but with the aluminum collar Tresse removed. The previous stipulations were the same for Unterarzt.
3. The Unterführeranwärter wore on the bases of his shoulder straps, a 0.9 cm wide aluminum braid.
4. Orders were issued for a series of command flags to be introduced but it is highly doubtful that they were put into service before the war's end. They were to be carried on the left fender of vehicles assigned to "Transportkorps Speer."

Uniform Accessories

Belts and Buckles

For Officers

By an order dated May 20, 1935, the wearing of the Luftsportverband style buckle (an NSDAP eagle within a wreath of laurel leaves) was to be discontinued and replaced by a newly designed buckle for the Luftwaffe. It was in an oval shape, of silver color and with a gold Luftwaffe eagle for officers and officials. For general ranks, it was also oval but of a gold color with a silver Luftwaffe eagle.[1] This buckle was worn with either an aluminum brocade belt or a light-brown belt with cross-over strap.* The pebbled double-claw belt buckle was also worn extensively by officers.

NOTE:
1. *Shortly after the outbreak of hostilities in 1939, the wearing of the cross-over strap for leather belts and the aluminum brocade belt was discontinued for the duration of the war.
2. Luftwaffe officers wore leather trimmings of a light-brown color, whereas NCOs and men wore dark-brown leather belts and equipment.

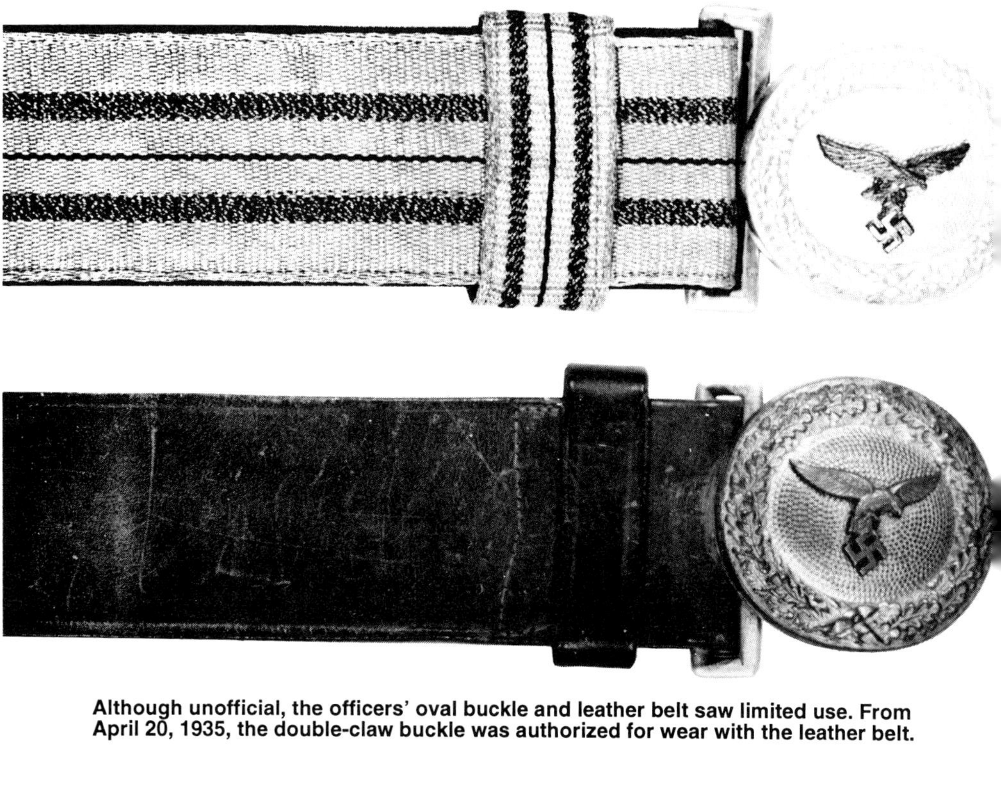

Although unofficial, the officers' oval buckle and leather belt saw limited use. From April 20, 1935, the double-claw buckle was authorized for wear with the leather belt.

[1] Luftwaffen-Verordnungsblatt, June 3, 1935. The earliest availability date for the newly designed belt buckle was mid-July, 1935.

A NCO candidate from the 1st Parachute Division wearing the dark brown belt and rectangular buckle.

A Luftwaffe officer in winter-issue parka wearing the double-claw buckle and belt.

For NCOs and Men

By the same order, the dark-brown belt with cross-over strap for NCOs and men (carried over from Luftsportverband regulations) was to have a newly designed buckle. The buckle with the NSDAP eagle was discontinued and in its place, a newly designed rectangular buckle with a Luftwaffe eagle within a wreath of laurel leaves, was introduced. Pre-war buckles were generally in a blue-grey color while those produced during the war were of an unpainted aluminum. The wearing of the cross-over strap for NCOs and men was ordered discontinued in early 1936. In its place were to be two cartridge pouches on either side of the belt buckle.[2]

[2] Luftwaffen-Verordnungsblatt, February 3, 1936. The order was numbered 119 and dated January 24, 1936.

Sidearms

Flyers' Dagger

The "Fliegerdolch" or Flyer's Dagger, which had been previously worn by the DLV, was adopted by the new Luftwaffe in 1935 and worn by officers, NCOs with portepee and officer candidates. Whereas the previously mentioned were permitted to wear the dagger with portepee, pilots, observers, wireless operators and flight mechanics of NCO and lower rank plus those NCOs not yet permitted the portepee, but awarded a flying badge, could wear the Flyer's Dagger without portepee.[3]

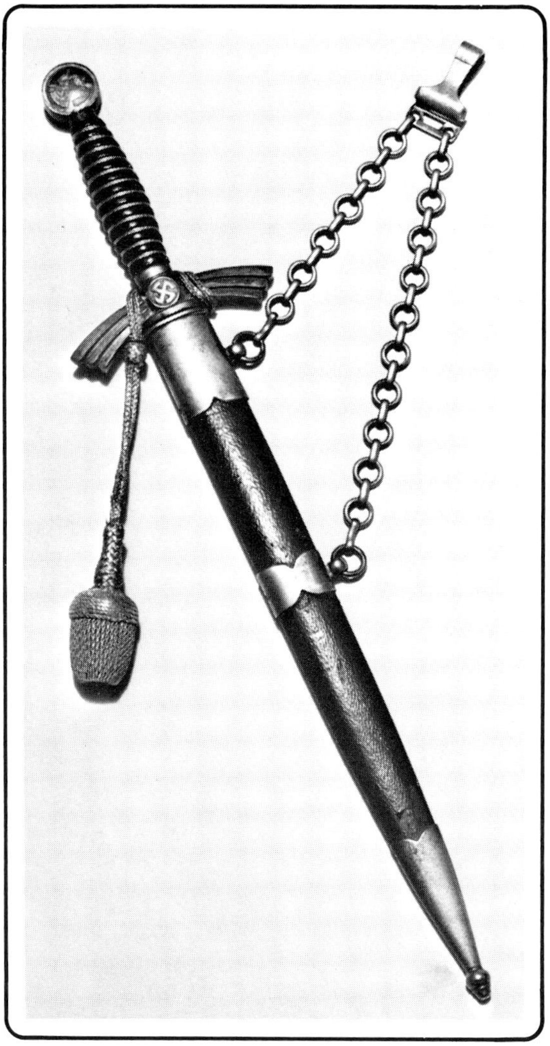

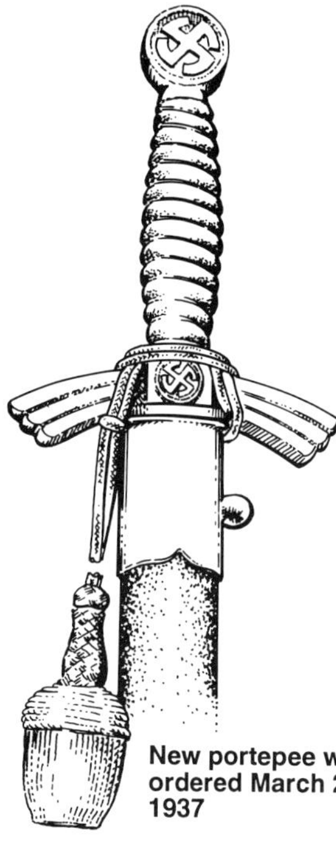

New portepee wrap ordered March 24, 1937

The Flyers' Dagger was initially produced with nickel-silver fittings (later polished aluminum). The swastika on both sides of the pommel and crossguard were gold colored and the handle and scabbard were covered with blue morocco leather. Note portepee wrap.

[3] Luftwaffen-Verordnungsblatt, March 30, 1936. The order is numbered 401 and dated March 27, 1936.

Officers Dagger

Officers plus officials and engineers of officer rank were to replace their Flyers' Dagger with the newly designed "Offizierdolch" or Officers' Dagger by October 1, 1937. Oberfähnriche, Unterärzte, Oberfeuerwerker and official candidates (Beamtenanwärter) in the rank of Oberfeldwebel, who had passed their officer examinations, were also permitted to wear the Officers' Dagger.[4] By an order dated March 12, 1940, the wearing of the Officers' Dagger was extended to all NCOs with portepee, from Feldwebel (or Wachtmeister) upwards. And also from this date, the Flyers' Dagger was to be worn exclusively by all NCOs and men (including Fähnrich) of the flying crews of the Luftwaffe.[5]

Regulation portepee wrap for officers' dagger

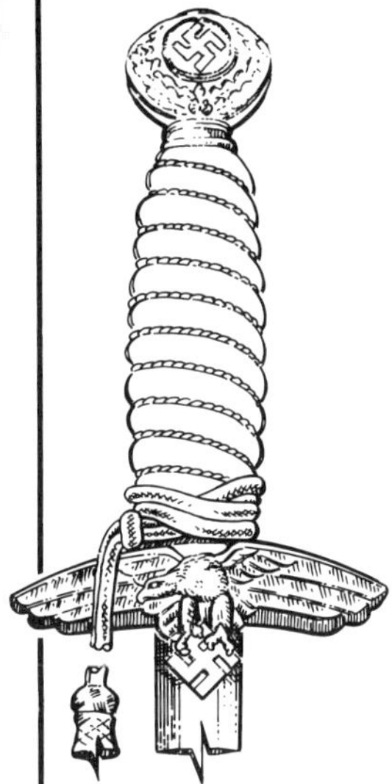

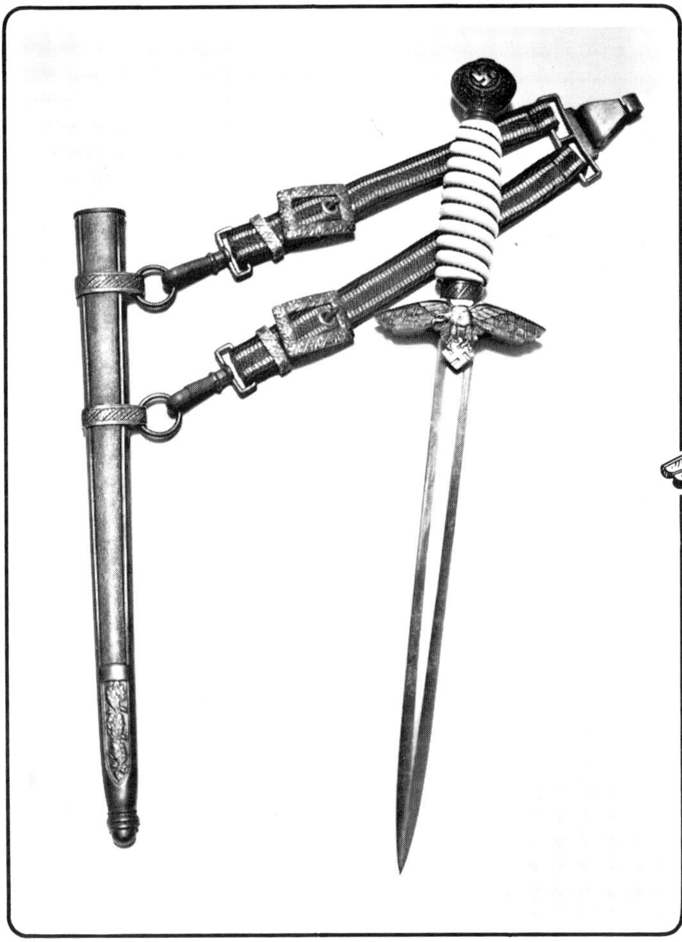

The officers' dagger was produced in cast aluminum with grips of ivory, plastic or celluloid covered wood (grip colors had no significance). Note detachable double strap hangers with dull aluminum fittings (after May 1942, general officer ranks wore gold finished fittings.

[4] Luftwaffen-Verordnungsblatt, July 19, 1937. The order is numbered 916.

[5] Hettler, E., *Uniformen der Deutschen Wehrmacht, Nachtrag 1939/1940.* p. XLV.

Pistols

In 1942, it was required to purchase a pistol for wear or wear an issue weapon. The available issue weapons from supply houses were as follows:[6]

German Manufacture:	Foreign Manufacture:
7.65 Mauser, Sauer & Sohn, Walther, Schmeisser	7.65 belg. Browning M37
9 mm Pistole 08 (luger and Pistole 38	9 mm Pist. belg. Armeemodell Pist. M24(t)

With the enemy advancing on all fronts, officers and officials of officer rank were to wear in public, for the duration of the war, a pistol in place of the dagger or sword.[7]

NOTES:
1. The wearing of a pistol with a caliber of 6.35 mm or smaller was forbidden (this order was not enforced within the recognized borders of Germany).
2. The pistol's model, caliber and serial number were to be entered into the wearer's Soldbuch.

Swords

Flyers' Sword

The Luftwaffe Flyer's Sword, which was introduced in 1934, was worn by officers and NCOs with portepee (not including Fähnriche). When worn with the Field Dress, it was attached to a blue leather frog which was suspended from the left pocket of the tunic. The sword was initially produced with nickel-silver fittings (later polished aluminum) with the swastika on both sides of the pommel and crossguard in gilt. The wire-wrapped handle and scabbard were covered in blue morocco leather. It was worn with the brocade belt (not NCOs), while in Service Dress, Reporting Uniform and Parade Dress. No portepee was worn with the Flyers' Sword.

Generals' Stichdegen

General officers did not wear the Flyer's Sword (Fliegerschwert) after the introduction of the specially designed "Degen." There were two patterns of the "Degen" or sword, the first had the Luftwaffe eagle on a shell guard and the second had the Luftwaffe eagle on the wire-bound grip. The grip was of white ivory on deluxe models and white celluloid on standard models. The hilt was in gilt.

[6] Luftwaffen-Verordnungsblatt, April 24, 1944. The order is numbered 528 and dated April 6, 1944.

[7] Luftwaffen-Verordnungsblatt, January 15, 1945. The order is numbered 42 and dated December 23, 1944.

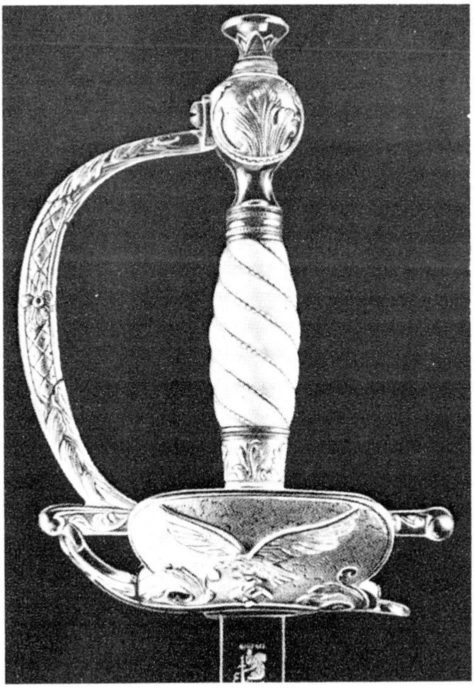

1st model Generals' "Degen"

2nd model Generals' "Degen"

Flyers' sword in detail

NOTES:
1. Daggers, swords and Degen were also worn by officials and bandmasters in equivalent ranks.
2. The Stichdegen was worn with the undress, walking-out dress and informal full-dress (day) uniforms. The flyers' sword was not worn.

The M84/98 Bayonet

Luftwaffe NCOs and enlisted men were issued the M84/98 bayonet (with dark brown frog) which was worn at all times when the individual was in uniform. It was in a blued finish with either bakelite or walnut grips and a 25 cm blade. The scabbard was of blued steel.

An optional dress sidearm was also available for wear during off-duty hours, with the Walking-Out Dress. The hilt and blade were normally nickel-plated, with black plastic, checkered grips and with a black steel scabbard. Two blade lengths exist in this dress version: a 20 cm blade length for NCOs and a 25 sm blade length for enlisted men.

In both cases, a distinctively colored bayonet knot (Faustriemen) was worn with all variations of dress (except the Field and Service Dress). The knot's color combination indicated the wearer's Staffel, battery or company.

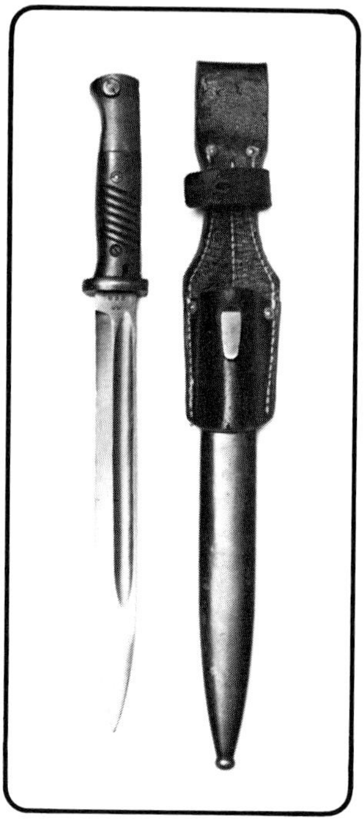

The M84/98 Bayonet

NCOs - The crown was dark-green, entwined with aluminum thread and the tassel was in aluminum cord. The strap and slide were in blue-grey.

Enlisted men - The strap was in blue-grey and the crown was in the appropriate Staffel, battery or company color. For slide and tassel colors, see following slide and tassel text.

The following colors indicate:

Dark-Green	Staff	as for the Army
White	1st unit	as for the Army
Red	2nd unit	as for the Army
Yellow	3rd unit	as for the Army
Cornflower Blue	4th unit	
Golden-Brown	5th unit	
Pink	6th unit	

Note NCO's bayonet knot (front rank) and that of the enlisted man (second rank).

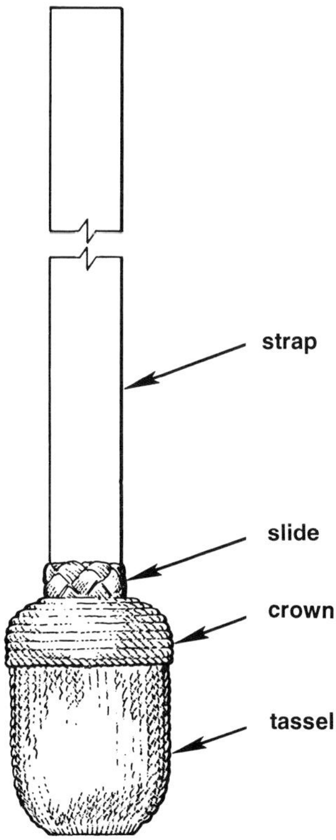

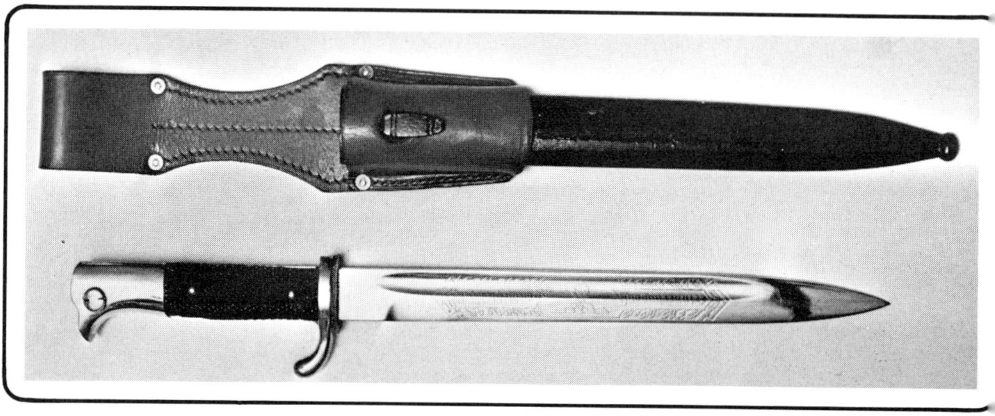

Model 98 pattern dress sidearm with etched blade

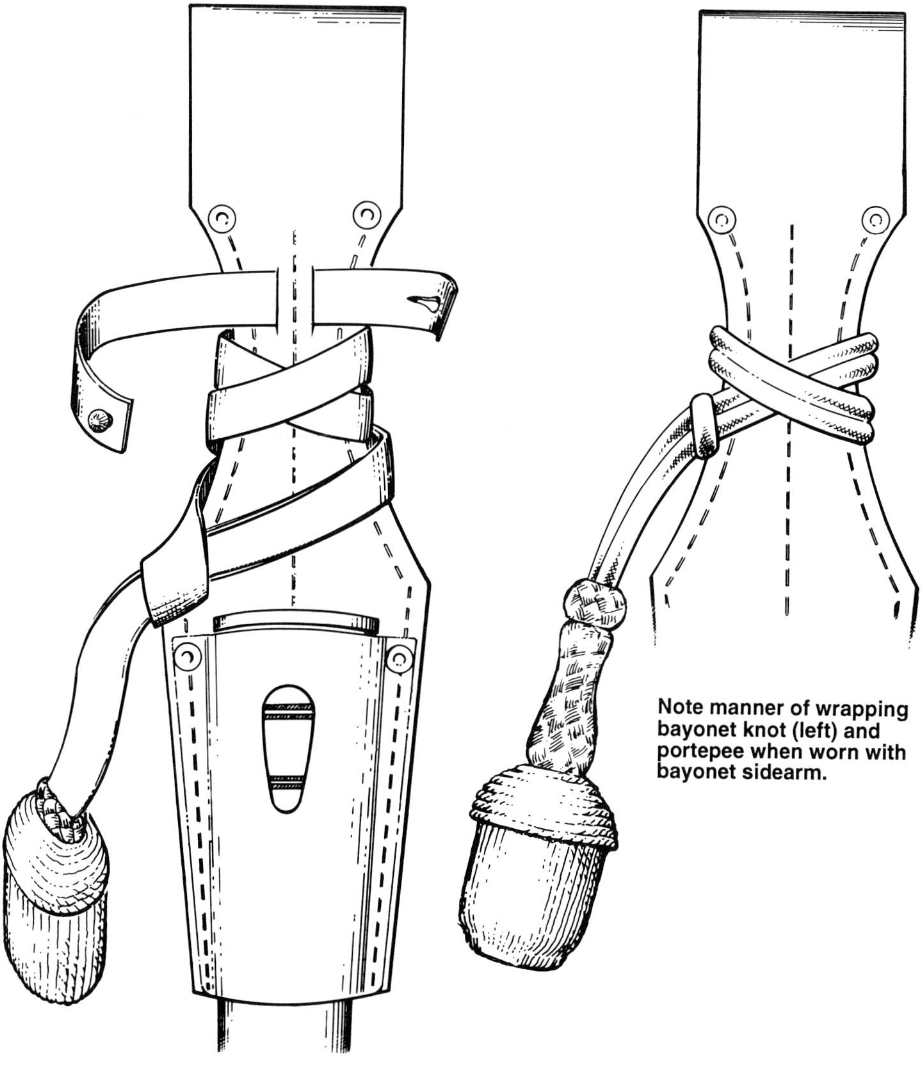

Note manner of wrapping bayonet knot (left) and portepee when worn with bayonet sidearm.

The Slide indicated:

the regimental staff or battalion (i.e. white - 1st battalion, red - 2nd battalion, yellow - 3rd battalion and cornflower-blue - IV battalion).

The Crown indicated:

the Staffel, battery, company or their staff within a Geschwader, regiment or battalion (i.e. white slide and red crown - 2nd company, 1st battalion).

The Tassel, which indicated a Gruppe or battalion staff or Staffel, battery or company formation, was:

(a) blue-grey for flying troop Staffel, Flak artillery battery and signals company personnel

(b) dark-green for Geschwader, regiment and battalion staffs, as well as for HQ Staffel, HQ battery, quartermaster, command company and support staffs.

Specific individual wear was as follows:[8]

(1) Air base command unit, training center unit, antiaircraft gunnery range, air munitions school, and armory and storage unit personnel wore the colors of the staff of the IV battalion: cornflower-blue slide, dark-green crown and dark-green tassel.

(2) Higher staffs, regimental staffs, independent battalion staffs and school staffs wore the bayonet knot with dark-green slide, crown and tassel.

(3) HQ Staffel, HQ battery, administrative, headquarters company and support unit staffs wore the staff bayonet knot with the crown in the successive colors of that unit, beginning with white.

(4) Training and course instruction units wore the bayonet knot of the appropriate Staffel, battery or company, in the successive colors of the unit, beginning with white.

(5) Air district medical battalion personnel wore the bayonet knot of the 1st company: white slide, white crown and blue-grey tassel.

(6) Replacement units wore the bayonet knot with dark-green slide, and the crown color determined in accordance with existing replacement units.

(7) Antiaircraft (machine-gun) reserve units wore a bayonet knot with a red slide, crown and tassel.

(8) Officer candidates in the rank of Gefreiter, Unteroffizier, Unterfeldwebel or Unterwachtmeister wore the NCOs' bayonet knot with blue-grey strap, blue-grey slide, green with entwined aluminum thread crown and aluminum tassel.

(9) Ranks from officer candidate of NCO rank (Fähnrich) and upwards wore with the dagger, bayonet and Degen (sword for generals), on all variations of dress, the aluminum portepee. The Flyer's Sword was worn without portepee.

[8] Hettler, E., *Uniformen der Deutschen Wehrmacht*, pp.109/110.

Aiguillettes

The aiguillette below was worn by officers, and officials, engineers and navigation experts of officer rank, plus musical directors and all band-masters. They were in an aluminum color for ranks up to Oberst, and in gold for generals. Note that general ranks utilized the style of aiguillette as that of naval admirals. The aiguillette was worn from the right shoulder with the Parade Dress, and the Day and Evening Dress.[9]

Hauptmann Mölders and his men in 1939.

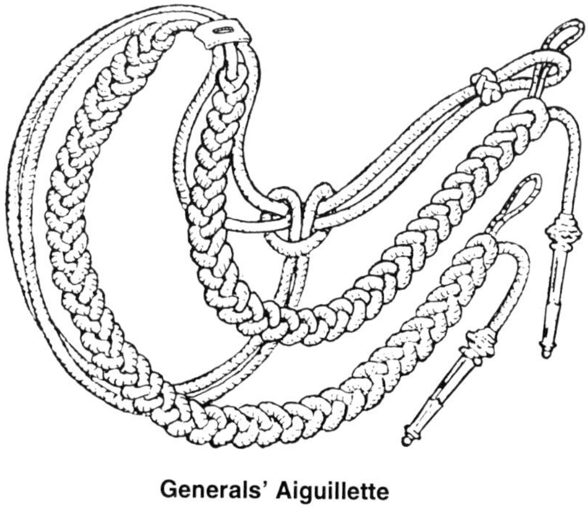

Generals' Aiguillette

[9] Hettler, *Uniformen der Deutschen Wehrmacht*, p. 109.

The Adjutants' Aiguillette

The adjutant aiguillette was also in aluminum and worn with the Field Dress, Service Dress, Reporting Uniform and Parade Dress (also on the great-coat).

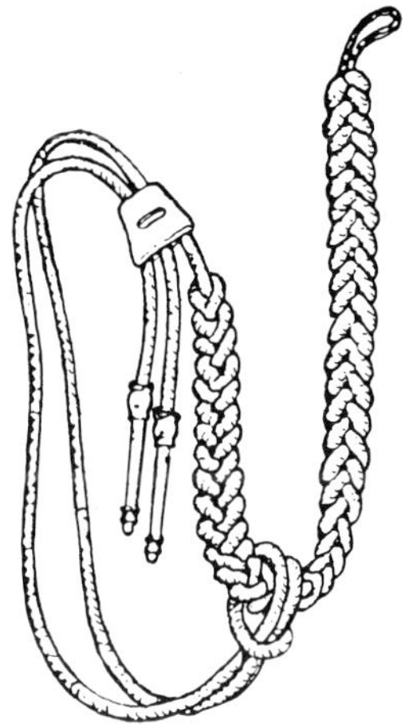

The Service Lanyard

The service badge or lanyard, in yellow wool, was worn by officers and NCOs of the day while engaged in service as transport officers and NCOs, dispatch officers, guard commanders of telegraph installations, aircraft directional radio operators, etc.

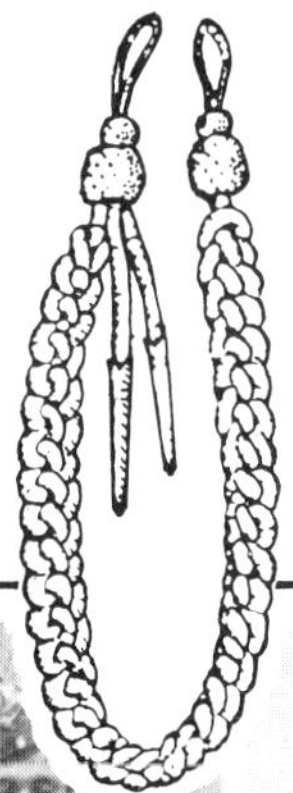

Recruits being led to their barracks after the issuance of clothing (note NCO candidate as NCO of the day with service lanyard).

Luftwaffe Specialty Badges

NCO's and men who were in a specialized field, were issued specialty badges which were worn on the left sleeve (1 cm above the cuff on the service tunic and at the same approximate location on the flying blouse). The badges were in matte-grey embroidery on a blue-grey backing. In place of the standard issue, NCOs and men could purchase matte-grey, hand-embroidered badges for wear.[1]

It has been attempted to cover the badges in their general sequence of initiation and issuance, therefore, the first grouping below was initiated from March 1935 - April 1937.[2]

1. Flying Personnel, except those entitled to wear the Pilot, Observer or Radio Operator's metal badge (Fliegendes Personal)
2. Flying Technical Personnel (Fliegertechnisches Personal)
3. Administrative NCO (Verwaltungsunteroffizier)
4. Mechanized Equipment Administrator (Kraftfahrzeuggerät)
5. Aircraft Equipment Administrator (Geräteverwalter für Flugzueggerät)
6. Searchlight Equipment Administrator (Geräteverwalter für Scheinwerfergerät)
7. Signals Equipment Administrator (Luftnachrichtengerät)
8. Qualified Telephone Operator (geprüfter Fernsprecher)
9. Qualified Telephone Operator, NCO with 2 mm matte-grey cord border (geprüfter Fernsprechunteroffizier)
10. Qualified Teletype Operator (geprüfter Fernschreiber)
11. Qualified Teletype Operator, NCO with 2 mm matte-grey cord border (geprüfter Fernschreibunteroffizier)
12. Qualified Radio Operator (geprüfter Funker)
13. Qualified Radio Operator, NCO with 2 mm matte-grey cord border (geprüfter Funkunteroffizier)
14. Signals Personnel for Flying and Flak Branches (Truppennachrichtenpersonal d. Fl. Truppe u. Flak)
15. Ordnance Personnel (Feuerwerker)
16. Armorer NCO, Flying and Signals Branch (Waffenunteroffizier für Fl. Truppe u. Ln. Truppe)
17. Armorer NCO, Flak Artillery and Regiment Göring (Waffenunteroffizier für Flakartillerie u. Regt. Göring)
18. Medical Personnel (Sanitätspersonal)
19. Motor Vehicle Driver (Kraftfahrer)
20. Flak Artillery Personnel, after 9 months service (Flakartilleriepersonal)
21. Military Boat Personnel, tenders, crash boats, etc. (Seemännisches Bootspersonal)

[1] Luftwaffen-Verorgnungsblatt, November 18, 1935, Order #860.

[2] Anzugordnung für die Luftwaffe, November 27, 1935, revised April 1, 1937, pp. 130-135.

1

2

3

4

5

6

An administrative NCO explaining to students the duties of a radio directional operator.

7

8

9

10

11

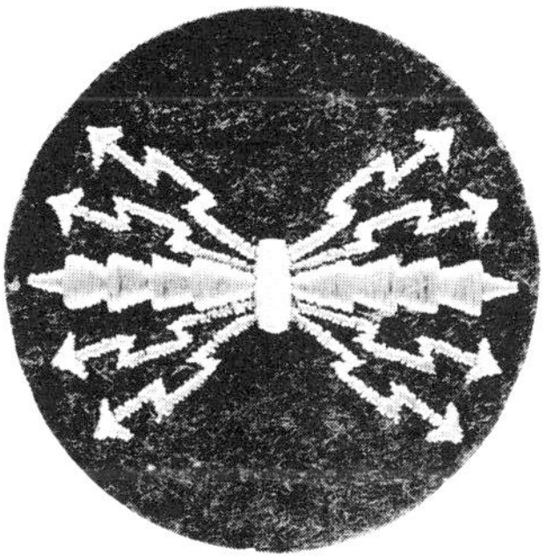

12

13

14

15

16

17

18

19

20

21

A Luftwaffe gun crew in readiness as British bombers approach the city.

Late 1938:
22. Qualified Directional Radio Operator (geprüfter Peilfunker)
23. Qualified Directional Radio Operator, NCO with 2 mm matte-grey cord border (geprüfter Peilfunkunteroffizier)
24. Qualified Sound Location Operator (geprüfter Horchfunker)
25. Qualified Sound Location Operator, NCO with 2 mm matte-grey cord border (geprüfter Horchfunkunteroffizier)

May 9, 1941:[3]
26. Graduate from a Technical Flight Preparatory School (ehemaliger Militärschüler der Fl. Techn. Vorschulen) (upper left sleeve)

August 12, 1942:[4]
27. Student at a Luftwaffe NCO School (Unteroffizierschüler der Luftwaffe) (right lower sleeve)

September 11, 1942:[5]
28. Master Radioman, with 2 mm matte-grey cord border (Funkmeister)
29. Aircraft Radioman, with certificate B (Flugzeugfunkwart mit Prüfung B)
30. Aircraft Radioman, with certificate C, with 2 mm matte-grey cord border (Flugzuegfunkwart mit Prüfung C)

October 19, 1942:[6]
31. Farrier (Hufbeschlagpersonal), from 8.9.-19.10.42 insignia in yellow (lower right sleeve)

Date of initiation unknown but presumably after 1942:
32. Air Raid Warning Personnel (Flugmeldepersonal)
33. Flak Crew, Loader, Fuse-Setter, etc. (Geschützbedienung, Ladekanonier, Zündereinsteller)
34. Transport Sergeant (Schirrmeister) (Order #1747, dated June 16, 1942)
35. Technical Sergeant (Technischer Unteroffizier)
36. Aerial Bomb Armorer (Waffenunteroffizier für leichte Bomben)
37. Heavy Aerial Bomb Armorer (Waffenunteroffizier für schwere Bomben)
38. Rangefinder (Entfernungsmesser)
39. Rangefinder, with 3 mm gold border for 1 year service (Entfernungsmesser) (Order #923, dated April 2, 1942)
40. Sound Location Operator (Horcher) (Order #923, dated April 2, 1942)
41. Sound Location Operator, with 3 mm gold border for 1 year service (Horcher) (Order #923, dated April 2, 1942)

---

[3] Luftwaffen-Verordnungsblatt, May 26, 1941, Order #540.

[4] Luftwaffen-Verordnungsblatt, August 24, 1942, Order #2121.

[5] Luftwaffen-Verordnungsblatt, September 21, 1942, Order #2391.

[6] Luftwaffen-Verordnungsblatt, October 26, 1942. This order was numbered 2673 and dated October 19, 1942. This badge was worn by Army and civilian blacksmiths on duty in the Luftwaffe as teachers and instructors. NCOs had a 2 mm matte-grey cord border.

22

23

24

25

26

27

28

29

30

31

32

33

34

35

36

37

38

39

40

41

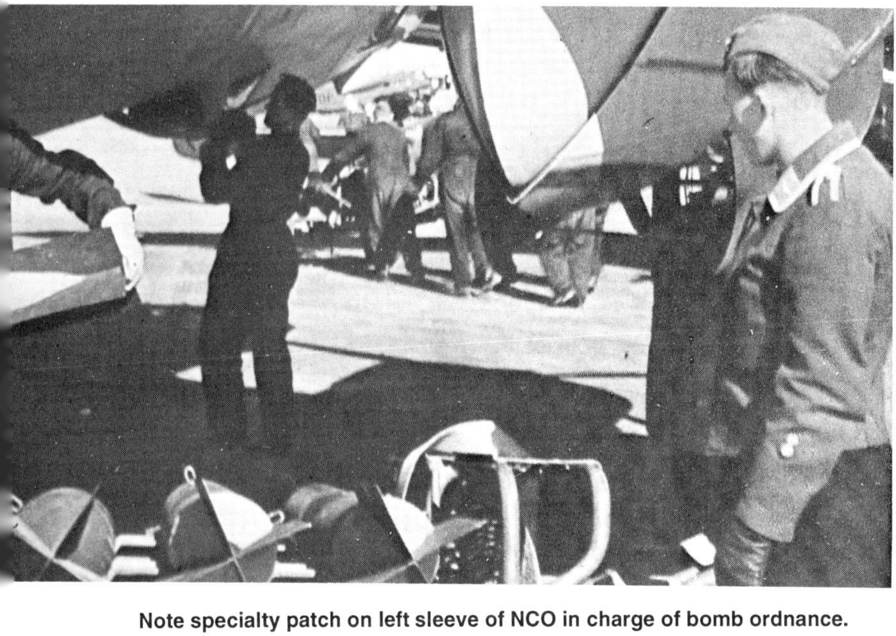

Note specialty patch on left sleeve of NCO in charge of bomb ordnance.

NOTE:

During the period between the passing of the specialty tests and the acceptance into a unit, the following specialists wore a strip of aluminum tresse below their badges. This tresse indicated a candidate specialist.

(Specialists permitted candidate insignia: 3, 4, 5, 6, 7, 8, 10, 12, 14, 15, 16, 17, 18, 19, 22, 24, 29, 31-38, and 40.

Generalmajor Ramcke

Badges, Clasps and Decorations

Flyers' Badges

Aircrew Badge (Fliegerschaftsabzeichen)

This badge was initiated on January 19, 1935 and initially worn by members of the Deutscher Luftsport-Verband who held a flying license. When the new Luftwaffe adopted it in early 1935, it was worn on the left breast pocket by pilots and observers. The horizontal oval wreath was in silver with a black eagle holding a silver swastika.

NOTE:
The following badges which had embroidered versions, were in aluminum or silver wire (where specified) for officers and the appropriate colored cotton thread for NCOs and enlisted men. It should be noted that these were purchased privately and not an awarded item.

Pilots' Badge (Flugzeugführerabzeichen)

On March 26, 1936, this newly designed badge was introduced and replaced the Aircrew Badge. The oval wreath was in silver with the raised portions of the design being polished. The eagle and swastika were either in a tarnished silver or a black finish. An embroidered version of this badge existed with the wreath in silver wire, the eagle in a dark-grey thread and the swastika in a matte-aluminum embroidery (officers').

The Pilot's Badge was worn in the middle of the left breast pocket of the tunic or in the approximate area of the Flying Blouse and Evening Full-Dress jacket. If the individual had been awarded the Iron Cross 1st Class, the badge was worn below it.[1]

[1] Anzugordnung für die Luftwaffe, revised April 1, 1937, p. 148.

Aircrew badge　　Pilots' badge

Luftwaffe Pilots' Badge (Civilian)
(Ziviles Flugzeugführerabzeichen der Luftwaffe)

Civilian pilots in the Luftwaffe were also permitted a pilot's badge...it was a pair of enameled blue wings with a slanted swastika in the middle. The wings were approximately 8 cm wide and 1.6 cm high. The detail of the wings and feathers, the outline of the insignia and the swastika were in gold. It was also available in an embroidered form.[2]

Observers' Badge (Beobachterabzeichen)

The Observer's Badge, also introduced on March 26, 1936, was for observers, assistant observers, navigators and bombardiers. The oval wreath was in silver with the raised portions of the design being polished. The eagle, in an observing position, was either in a tarnished silver or black finish. An embroidered version of this badge existed with the wreath in silver wire, the eagle in a dark-grey thread and the swastika in a matte-aluminum wire, on a blue-grey base (officers').[3]

Combined Pilot/Observer Badge
(Gemeinsames Flugzeugführer-und Beobachterabzeichen)

This badge was initiated on March 26, 1936, and was for those who held the pilot's and observer's certificates for a minimum of one year. The design of the badge is identical to that of the pilot's but with a gilt wreath, with the raised portions being polished, and a silver eagle and swastika. The embroidered version had the wreath in gold thread, the eagle in aluminum wire and the swastika in silver wire (officers').[4]

[2] Luftwaffen-Verordnungsblatt, April 1936. The order was numbered 420. (For illustration, see "Flying Civilian Employees," p. 251)

[3] Anzugordnung für die Luftwaffe, revised April 1, 1937, p. 148.

[4] Ibid, p. 149.

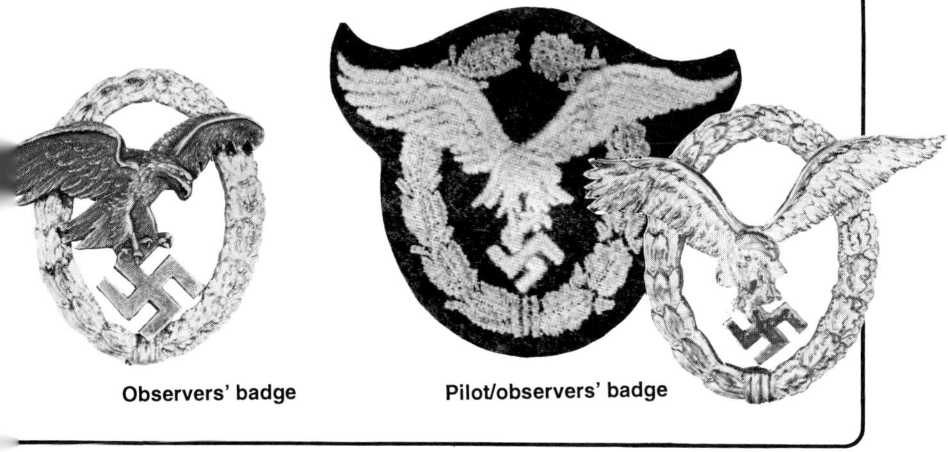

Observers' badge

Pilot/observers' badge

Air-gunner's badge/Wireless-Operator (Fliegerschützen-(Bordfunker) Abzeichen)

This badge for Air-Gunner/Wireless-Operators was also instituted on March 26, 1936. The wreath was in silver with the raised portions of the design plus the swastika being polished. The eagle, holding the lightning flashes (symbolic of wireless), was in tarnished silver or black. The embroidered version had the wreath in aluminum wire and the eagle and lightning flashes in a grey thread (officers').[5]

Air Gunner's and Flight Engineer's Badge (Fliegerschützen-(Bordschützen-und Bordmechaniker) Abzeichen)

The Air Gunner's and Flight Engineer's Badge was instituted on June 22, 1942 and was identical in design to the Air-Gunner/Wireless-Operator's Badge with the exception of the lightning flashes being omitted. It was worn by all air gunners (except combination wireless-operator/air gunners) and flight engineers and flight meteorologists.

On April 25, 1944, a variation of the Air Gunners and Flight Engineer's Badge was introduced for unqualified air gunners without certificate. It was identical in design but with a black wreath and silver eagle (no lightning flashes). It was awarded for a minimum of ten combat flights. This number could be reduced if the unqualified air gunner was wounded during one of the flights.[6]

Flyer's Commemorative Badge (Fliegererinnerungs Abzeichen)

This badge was awarded to all flying personnel (including parachute troops) who had been relieved of all air crew duties, World War I pilots with at least four years service, plus other personnel with at least fifteen years service time. This commemorative badge was also extended to those crippled by an air-associated accident and to the next-of-kin in case of a flying accident resulting in death. The oakleave wreath was in silver with a polished swastika and the eagle sitting on a boulder was in tarnished silver.

[5] Ibid, p. 149.

[6] Luftwaffen-Verordnungsblatt, May 8, 1944. The order was numbered 595 and dated April 25, 1944.

An embroidered version of the commemorative badge was also available, but only for officers. The wreath and swastika was in aluminum wire and the eagle and boulder was in a grey thread.

The metal version of this badge could be worn with civilian dress. A stickpin form was also permitted on the left lapel. Disabled veterans were also allowed to wear a miniature air crew badge with their civilian dress.[7]

Parachutist Badge (Fallschirmschützen Abzeichen)

The Parachutist Badge was initiated on November 5, 1936, for qualified parachutists in the Luftwaffe. The wreath was black with a gold diving eagle and swastika (the highlights of the eagle and swastika were polished). The embroidered version has the wreath in white thread for NCOs and men and in aluminum wire for officers. The eagle and swastika was in gold colored thread for NCOs and men and in gold wire for officers.[8]

Army

The Army parachutist badge at left was instituted on June 15, 1937 and worn by Army parachutists until January 1, 1939, when they came under the command of the Luftwaffe. The Army eagle and wreath were in gilt and the diving eagle in silver.

Luftwaffe

NOTE:

An order dated October 19, 1939, gave the following regulations for the awarding of Flier's Badges for the duration of the war:[9]

1. Pilot's Badge - after obtaining the pilot's certificate
2. Observer's Badge - after five combat flights or two months after obtaining the Observer's, Navigator's or Bombardier's certificate
3. Wireless-Operator/Flight Engineer/Air Gunner - after five combat flights or two months after obtaining the appropriate certificate

[7] Anzugordnung für die Luftwaffe, revised April 1, 1937, pp.149/150.

[8] Luftwaffen-Verordnungsblatt, November 16, 1936. The order was numbered 1491 and dated November 5, 1936.

[9] Luftwaffen-Verordnungsblatt, November 6, 1939. The order was numbered 314 and dated October 19, 1939.

Glider Pilots' Badge (LW-Segelflugzeugführerabzeichen)

The Glider Pilot's Badge was instituted on December 16, 1940, and awarded to those having obtained the military glider pilot's certificate. The wreath and swastika were in silver and the eagle was black or in tarnished silver. It could be worn only on a Wehrmacht or NSFK uniform. The embroidered version has the wreath in aluminum wire, the swastika in bright (silver) wire, and the eagle in grey for officers.[10]

[10] Luftwaffen-Verordnungsblatt, December 30, 1940. The order was numbered 1635 and dated December 16, 1940.

War Badges

Luftwaffe Flak Badge (Flak-Kampfabzeichen der Luftwaffe)

This badge, which was instituted on January 10, 1941, was in silver or white metal and awarded upon completion of the following requirements:[11]

(a) for heavy batteries - at least five enemy aircraft downed

(b) for light batteries - at least five enemy aircraft downed

(c) for searchlight batteries - if through their detection, a minimum of fifteen engagements is initiated (either by flak or night fighters) or if through their detection five enemy aircraft were destroyed

(d) additionally, if in three successful engagements against land (tanks, bunkers, etc.) or sea (ships) targets

(e) for battery commanders, if half of their command has been awarded the Flak Badge

NOTE:
An order dated January 6, 1945, permitted Flakwaffenhelferinnen, Flakhelferinnen and RAD.-Flakwaffenhelferinnen assigned to Flak and searchlight batteries to be awarded the Flak Badge.

Ground Combat Badge of the Luftwaffe (Erdkampfabzeichen der Luftaffe)

On March 31, 1942, Reichsmarschall Göring announced that a new Ground Combat Badge had been instituted for his many ground combat formations...whose issuance was to begin on April 1, 1942. The wreath and Luftwaffe eagle were in matte-silver with polished highlights. The clouds, lightning flash and ground were in tarnished silver or black (the outline of the lightning flash was polished bright). The embroidered version was in the same color scheme as described above. Regulations governing the awarding of the badge were as follows:[12]

(a) three separate actions during front-line duty (with infantry or artillery weapons), including hand-to-hand combat

(b) Luftwaffe members holding the Army's Infantry, General Assault or Tank Battle Badge (permitted worn to this date), were to exchange them for the Luftwaffe Ground Combat Badge

[11] Luftwaffen-Verordnungsblatt, April 13, 1942. The order was numbered 921 and dated March 31, 1942.

[12] Luftwaffen-Verordnungsblatt, November 20, 1944. The order was numbered 1574 and dated November 3, 1944.

(c) the badge was awarded to his next-of-kin if the individual was killed during his third combat action

(d) members of Sturmgeschütz units and parachutists were also eligible for the badge provided they had fulfilled their three engagement requirement on three separate days.

On November 10, 1944, higher grades of this badge were initiated for the Luftwaffe's ground combatants. They were basically the same design as the Ground Combat Badge only larger, plus small numbered boxes which were positioned on the base of the wreath (i.e. "25" (II.Klasse) for 25 engagements, "50" (III.Klasse) for 50 engagements, "75" (IV.Klasse) for 75 engagements and "100" (V.Klasse) for 100 engagements).[13]

Ground combat badge

Higher grade

Luftwaffe Tank Battle Badge (Panzerkampfabzeichen der Luftwaffe)

The Tank Battle Badge, instituted on November 3, 1944, was in two forms:[14]

1. silver wreath and matte-silver eagle, with the tank in black for:

(a) tank commanders, gunners, drivers and radiomen for a minimum of three combat engagements on three separate days,

(b) also, tank repair crews providing they were in the front-lines and engaged by the enemy on three occasions on three separate days,

(c) also, medical personnel attached to armored units which were engaged on three occasions on three separate days

2. black wreath and matte-silver eagle, with the tank in black for:

(a) Panzergrenadier formations who manned front-line positions and engaged in three actions on three separate days,

(b) also, medical personnel assigned to Panzergrenadier units engaged in at least three actions on three separate days,

(c) also, armored reconnaissance units if engaged in at least three actions on three separate days

[13] Ibid.

[14] Ibid.

On November 10, 1944, higher grades of the Luftwaffe's Tank Battle Badge were initiated. They were basically the same design as the I.Klasse badge, only larger, plus small numbered boxes which were positioned on the base of the wreath (i.e. "25" (II.Klasse) for 25 engagements, "50" (III.Klasse) for 50 engagements, "75" (IV.Klasse) for 75 engagements and "100" (V.Klasse) for 100 engagements).[15]

Luftwaffe Sea Battle Badge (Seekampfabzeichen der Luftwaffe)

By an order dated November 11, 1944, Göring initiated the Luftwaffe Sea Battle Badge which was to be awarded to all members of the Luftwaffe (soldiers, officials and civilian employees) who constituted crews of air/sea rescue launches, supply ships or other special Luftwaffe surface craft.[16] It was awarded to:

1. leaders and crews of supply ships and other special Luftwaffe surface craft for:
   (a) sixty days at sea in the North or East Sea, between 5 and 20 degrees longitude and south of the 60th degree of latitude or,
   (b) twenty days at sea in the Mediterranean, including the Aegean and Black Seas

[15] Ibid.

[16] Luftwaffen-Verordnungsblatt, January 8, 1945. The order was numbered 19 and dated November 27, 1944.

2. leaders and crews of air/sea rescue launches for:

(a) twenty sea days with a minimum of one rescue attempt or a three hour sea duration each day or,

(b) ten sea days (a successful air/sea rescue mission counted as two sea days)

NOTES:

1. A sea day consisted of a minimum of 10 hours at sea. Days with less than 10 hours at sea could be accumulated on an hourly basis (i.e. 14 accumulated hours consisted one sea day).

2. The Luftwaffe Sea Battle Badge could not be awarded for the same action which entitled the participant to one of the Kriegsmarine badges.

The badge consisted of a gold, oval oakleave wreath, an old-silver Luftwaffe eagle and a ship, smoke-stack, rigging and smoke in dark-grey. As indicated in a copy of the original order at right, some copies of this late-war badge did indeed exist, but most likely did not see actual issuance.

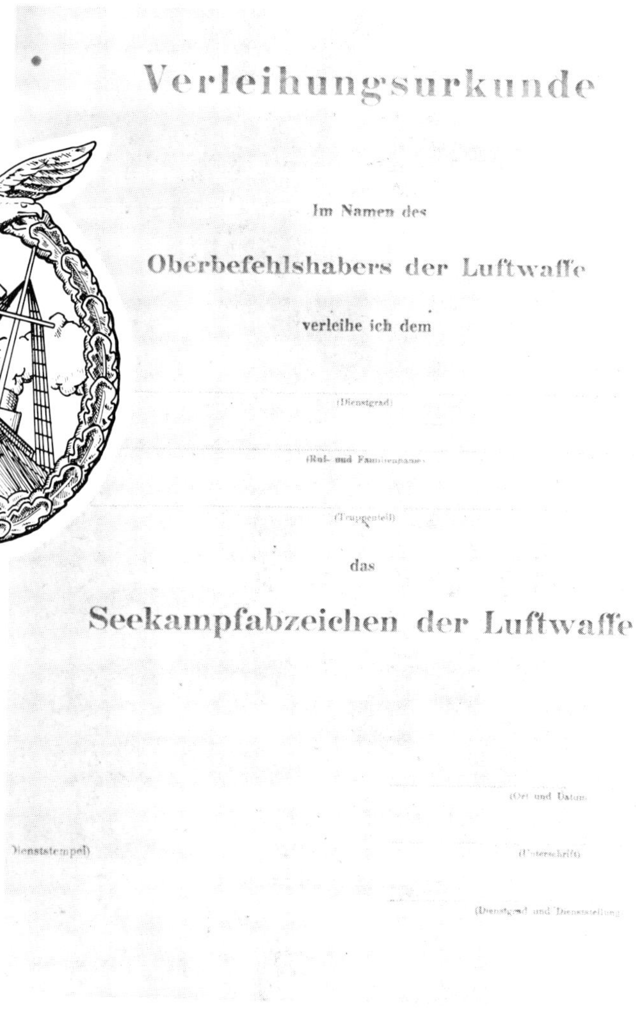

Verleihungsurkunde

Im Namen des

Oberbefehlshabers der Luftwaffe

verleihe ich dem

(Dienstgrad)

(Ruf- und Familienname)

(Truppenteil)

das

Seekampfabzeichen der Luftwaffe

(Ort und Datum)

(Dienststempel)

(Unterschrift)

(Dienstgrad und Dienststellung)

Award document for the Luftwaffe Sea Battle Badge

12 L. V. Bl. 1945 2. Ausgabe vom 8. Januar 1945

der für die Gaue jeweils festgesetzten Höchstzahlen.

5. Da von dem bevorzugten Verzicht auf den Widerruf nur in beschränktem Maße, und zwar jeweils zum 30. 1., 20. 4. und 9. 11. Gebrauch gemacht wird, ist damit zu rechnen, daß eingereichte Vorschläge längere Zeit zurückgestellt werden müssen.

6. Über die Vorlage von Anträgen sind die in Frage kommenden Soldaten nicht zu unterrichten, um Enttäuschungen bei etwaigen Ablehnungen zu vermeiden.

7. Bei Verzicht auf den Widerruf wird das Wehrmeldeamt und der Truppenteil von der Entscheidung verständigt und die Verzichturkunde dem Einheitsführer mit der Bitte übersandt, sie dem Soldaten in feierlicher Form auszuhändigen und den Aushändigungstag auf der Rückseite der Urkunde zu vermerken. Bei Aushändigung der Urkunde ist der Volkslistenausweis einzuziehen.

O. K. W., 3. 12. 1944
Az. 14 x 14 NSFW/4 (J) I Nr. 16630/44

O. K. L., 30. 12. 1944,
Az. 12 1 18 (Wehramt I 1 IA).

L. V. Bl. S. 11

Verordnungen und Befehle des R. d. L. und des O. K. L.

19. Stiftungsurkunde
»Seekampfabzeichen der Luftwaffe«

Um auch den Angehörigen des seemännischen Personals der Luftwaffe ein sichtbares Zeichen der Anerkennung für Bewährung im Einsatz als Besatzungen von Seenotfahrzeugen, Flugsicherungs-, Versorgungs- und anderen Spezialschiffen der Luftwaffe zukommen zu lassen, und in Würdigung der Tatsache, daß sie bei Durchführung ihrer schwierigen Aufgaben ständiger Bedrohung durch feindliche Luft- und Seestreitkräfte und bei Einsatz in minenverseuchten Gewässern und im Kampf mit den Elementen zusätzlichen Gefahren ausgesetzt sind, genehmige ich die Einführung des oben abgebildeten

»Seekampfabzeichens der Luftwaffe«.

Die Verleihung des »Seekampfabzeichens der Luftwaffe« übertrage ich den Oberbefehlshabern der Luftflotten und den ihnen gleichgestellten Befehlshabern. Die Verleihung hat nach den beifolgenden Bestimmungen zu erfolgen. Für genaues Einhalten dieser Verleihungsbestimmungen sind mir die zuständigen Vorgesetzten verantwortlich.

Hauptquartier des Ob. d. L., den 27. 11. 1944.

Der Reichsmarschall des Großdeutschen Reiches und Oberbefehlshabers der Luftwaffe
Göring

Original order for the Sea Battle Badge

Luftwaffe Close Combat Clasp (Nahkampfspange der Luftwaffe)

This clasp, initiated on November 3, 1944 by Göring, was for those Luftwaffe ground personnel who had participated in hand-to-hand combat. It was to be worn 1 cm above the medal ribbon bar (as for the Operational Flying Clasps) and was in three classes:[17]

(a) I.Klasse in bronze* - 15 close combat days
(b) II.Klasse in silver* - 30 close combat days
(c) III.Klasse in gold* - 50 close combat days

*In all three grades, the hand grenade, bayonet, and Luftwaffe eagle were in matte-silver. The remainder of the clasp was in its appropriate color.

[17] Luftwaffen-Verordnungsblatt, November 20, 1944. The order was numbered 1574 and dated November 3, 1944.

NOTES:

1. A special 21 day furlough was granted recipients of the gold Close Combat Clasp, excluding traveling time.[18]

2. The first awarding of the gold clasp was to an NCO by Hitler, who had not yet received the 1st class Iron Cross...at this time, both decorations were awarded him by the Führer. It was therefore recommended that for all future awards of the silver Close Combat Clasp, an individual should be considered eligible for the 1st class Iron Cross, if he had not already been awarded it. Furthermore, it was recognized that if an individual had proven himself during 50 days of close combat and was eligible for the golden Close Combat Clasp, he should also be recommended for the German Cross in gold.[19]

Besitzzeugnis

Im Namen des

Oberbefehlshabers der Luftwaffe

verleihe ich

dem ....

(Dienstgrad)

....

(Vor- und Familienname)

....

(Truppenteil)

für tapfere Teilnahme an .... **Nahkampftagen**

Die .... **Stufe der**

Nahkampfspange der Luftwaffe

....

(Ort und Datum) (Unterschrift)

(Dienststempel) ....

(Dienstgrad und Dienststellung

Award document for the Close Combat Clasp

[18] Luftwaffen-Verordnungsblatt, January 15, 1945. The order was numbered 61 and dated January 8, 1945.

[19] Luftwaffen-Verordnungsblatt, January 22, 1945. The order was numbered 79 and dated January 12, 1945.

808 L. V. Bl. 1944 49. Ausgabe vom 20. November 1944

Verordnungen und Befehle des R. d. L. und des O. K. L.

Der Reichsmarschall des Großdeutschen Reiches
und Oberbefehlshaber der Luftwaffe

1574. Verordnung über die Einführung einer Nahkampfspange und eines Panzerkampfabzeichens der Luftwaffe

Der zunehmende Einsatz der Luftwaffeneinheiten, insbesondere der Fallschirmtruppen, in den Erdkämpfen der Abwehrfronten und an den Brennpunkten des derzeitigen Kriegsgeschehens sowie die Art der Teilnahme an diesen Kämpfen gibt mir Veranlassung, die Einführung weiterer Kampfabzeichen für die Luftwaffe zu genehmigen.

In Anlehnung an die beim Heer eingeführten Kampfabzeichen stifte ich als sichtbares Zeichen der Anerkennung für tapferes Verhalten im Kampf mit der blanken Waffe und Nahkampfmitteln Mann gegen Mann, zugleich auch als Ansporn zu höchster Pflichterfüllung

die »Nahkampfspange der Luftwaffe«

und für Bewährung im Kampf beim Einsatz als Angehörige von Panzer-, Panzerspäh- und Panzergrenadiereinheiten der Luftwaffe

das »Panzerkampfabzeichen der Luftwaffe«.

Die Verleihung dieser Kampfabzeichen der Luftwaffe hat nach den beifolgenden Bestimmungen zu erfolgen. Für genauestes Einhalten der Verleihungsbestimmungen sind die zuständigen Vorgesetzten verantwortlich.

Durchführungsbestimmungen erläßt der Chef der Personellen Rüstung und Nationalsozialistischen Führung der Luftwaffe.

Hauptquartier des Ob. d. L., den 3. 11. 1944.

Göring

1575. Durchführungsbestimmungen zu der Verordnung über die Einführung einer Nahkampfspange und eines Panzerkampfabzeichens der Luftwaffe

I.

»Nahkampfspange der Luftwaffe«

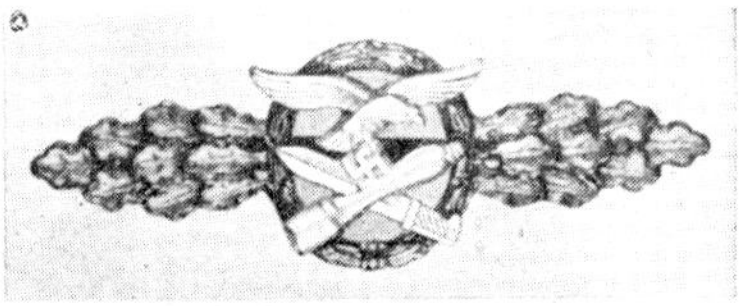

1. Die »Nahkampfspange der Luftwaffe« kann allen Angehörigen der Luftwaffe verliehen werden, die sich mehrfach im Nahkampf Mann gegen Mann mit der Waffe in der Hand ihrer Aufgabe entsprechend bewährt haben.

Das Erdkampfabzeichen der Luftwaffe kommt unabhängig hiervon bei Erfüllung der geforderten Bedingungen zur Verleihung.

2. Die »Nahkampfspange der Luftwaffe« wird in 3 Stufen verliehen, d. h. in Bronze, Silber und Gold. Sie ist ebenso wie die Frontflugspange 1 cm über der Ordensschnalle in und außer Dienst zu tragen. Bei Erwerb der höheren Stufe ist die vorhergehende abzulegen und zurückzugeben.

3. Die Bedingungen für die Verleihung sind:
für die 1. Stufe in Bronze 15 Nahkampftage
für die 2. Stufe in Silber . 30 Nahkampftage
für die 3. Stufe in Gold . . 50 Nahkampftage

4. Als Nahkampftage sind anzurechnen alle Kampftage, an denen die auszuzeichnenden Kämpfer Gelegenheit fanden, das »Weiße im Auge des Feindes« zu sehen, d. h. mit Nahkampfwaffen mit dem Gegner Mann gegen Mann im Kampf bis zur letzten Entscheidung zu stehen.

Nahkämpfe dieser Art können beim Großangriff, beim Spähtruppgang, in der Abwehr, bei einem einzelnen Meldegang, beim Zusammentreffen mit feindlichem Spähtrupp usw. gegeben sein. Der Ort, ob bei den Gefechtsvorposten, im Vorfeld, in der HKL, in der Artilleriefeuerstellung, im rückwärtigen Operationsgebiet (auch Bandengebiet) oder bei feindlichem Überfall auf eine Nachschubkolonne usw., ist belanglos.

5. Jeder Soldat, der ungeschützt und zu Fuß einen derartigen Nahkampf zu bestehen hatte und sich bewährt hat, ist Anwärter auf die Nahkampfspange.

Original order for the Luftwaffe Close Combat Clasp

Operational Flying Clasps

By an order dated January 30, 1941, a series of Operational Flying Clasps was initiated for flying personnel. They not only indicated the type of squadron to which an individual belonged, but the wearer's number of operational flights as well. Initially, the Flying Clasps encompassed several squadron functions such as:[20]

– Fighter, as well as Long-Range Day Fighter and Air-to-Ground squadrons

– Bomber, as well as Dive-Bomber, Transport and Glider squadrons

– Reconnaissance, as well as Air/Sea Rescue and Meteorological squadrons

Flying personnel eligible for these clasps were: pilots, observers, bombardiers, wireless-operators, flight mechanics, air gunners, war correspondents, as well as officials and members of the engineer and navigation corps. The number of operational flights was determined by the base color of the clasp:

(a) 20 operational flights - bronze
(b) 60 operational flights - silver
(c) 110 operational flights - gold

NOTE:
The emblems within the above clasps (all grades) were in tarnished silver, black or brown (crossed swords on the Air-to-Ground Clasp are in steel-blue).

An operational flight was counted if it penetrated a minimum of 30 km behind enemy lines or if the enemy was engaged. When over the sea, the flight had to exceed 100 km from the nearest friendly coast. This distance was reduced to 30 km if the flight entered an enemy's island or continental air-space.

Oberst Hajo Hermann wearing the Bomber Operational Clasp with pennant

[20] Luftwaffen-Verordnungsblatt, February 10, 1941. The order was numbered 137 and dated January 30, 1941.

Operational flights of over a four hour duration were counted as double, if half of this time was over the enemy's homeland (island or continent). Flights of over an eight hour duration were counted as triple. As was the case with Flier's Badges, Flight Clasps were also available in an embroidered version, but at the wearer's own expense. Officer's versions were in metal wire and those for NCOs and men were the appropriate colored cotton thread. Lost Operational Flying Clasps could be replaced (purchased) from the Verkaufsabteilung der Luftwaffe, Berlin SW 68, Puttkamerstrasse 16/18. The embroidered clasps were also available at this address.

Before the war's end, many specialized squadrons requested and were granted their own Operational Flying Clasp. The final results were as follows:

(a) For Day Fighters (Tagjäger), with the center design being a winged arrow pointing upwards. (initiated January 30, 1941)

NOTE:
Those Day Fighter Squadrons stationed near the Channel coast were permitted to count flights using the following additional regulations after January 28, 1943:[21]
(a) three ship escort flights with a minimum of 30 minutes duration each
(b) three fighter engagements with a minimum of 30 minutes duration each and a minimum distance of 50 km from his own coastline
(c) responding to three general alarms with a minimum of 30 minutes duration each (over the sea) and a minimum distance of 50 km from his own coastline.

(b) For Short-Range Night Fighters (Nahnachtjäger), with the center design being a winged arrow pointing upwards within a black wreath. (initiated October 14, 1942)

[21] Luftwaffen-Verordnungsblatt, February 8, 1943. The order was numbered 284 and dated January 28, 1943.

(c) For Long-Range Day Fighters and Air-to-Ground Support Squadrons (Zerstörer und Schlachtflieger), with the center design being a winged arrow pointing downwards (after April 12, 1944, this was for "Zerstörer" only). (initiated May 26, 1942)

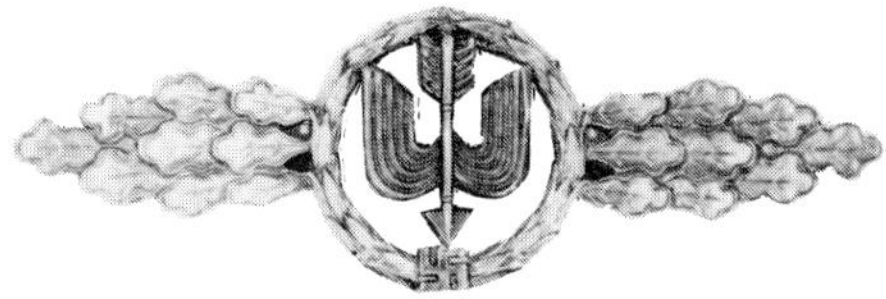

(d) For Long-Range Night Fighter and Night Intruder Squadrons (Fernnachtjäger), with the center design being a winged arrow pointing downwards within a black wreath. (initiated October 14, 1942)

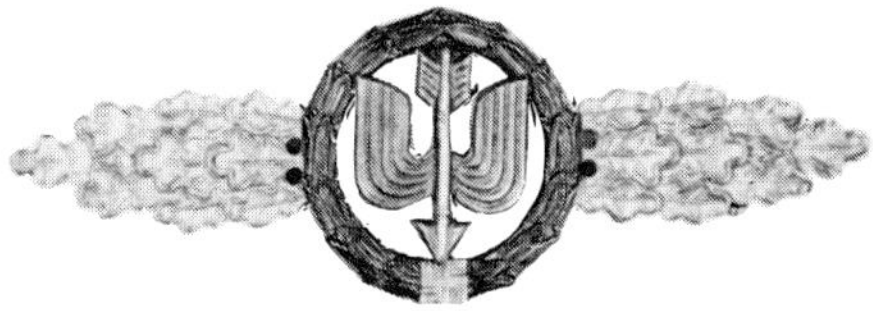

(e) For Heavy, Medium and Dive-Bombers (Kampf- und Sturzkampfflieger), with the center design being a winged bomb pointing downwards. (initiated January 30, 1941)

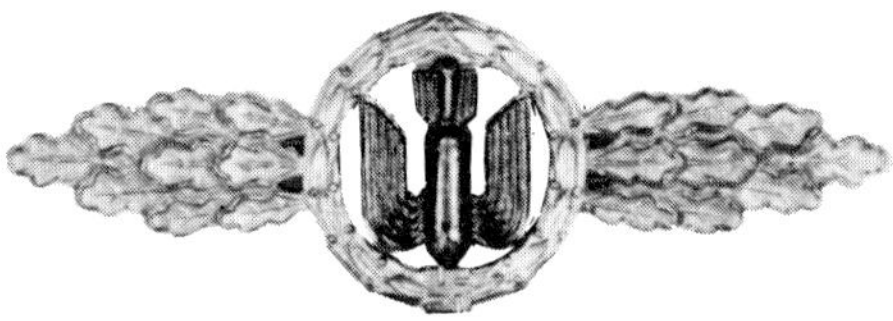

(f) For Reconnaissance, Air/Sea Rescue and Meteorological Squadrons (Aufklärer, Seenotflieger-Verbände und Wetter-Erkundungs-Staffeln), with the center design being an eagle's head, facing left. (initiated January 30, 1941)

NOTE:
Members of Sea-Mine Location units were also awarded the Reconnaissance Flying Clasp. Three sea-mine location flights constituted one operational flight, providing each flight duration was no less than 30 minutes. Flights were accumulated from January 1, 1944.[22]

[22] Luftwaffen-Verordnungsblatt, June 12, 1944. This order was numbered 781 and dated May 31, 1944.

(g) For Transport and Glider Squadrons (Transport und Luftlandeflieger), with the center design being the Luftwaffe flying eagle. (initiated November 19, 1941)

(h) For Air-to-Ground Support Squadrons (Schlachtflieger), with the center design being two crossed swords. (initiated April 12, 1944)

NOTE:
This clasp was awarded to Oberst Rudel in recognition of his 2,000 operational flights, in solid gold with diamonds and with a platinum wreath and crossed swords (April 1944).[23]

On June 26, 1942, a pendant (small gilt star flanked by laurel leaves...highlights polished) for the golden Operational Flying Clasp was initiated. It was added to the clasp and indicated the following:[24]

(a) Fighter and Transport Squadrons - 500 flights

(b) Dive-Bomber, Long-Range Day Fighter and Air-to-Ground Squadrons - 400 flights

(c) Bomber, Air/Sea Rescue and Meteorological Squadrons - 300 flights

(d) Reconnaissance and Night Fighter Squadrons - 250 flights

[23] Hans Ulrich Rudel, *Stuka Pilot* (New York: Ballantine Books, Inc., 1961). p.145.

[24] Luftwaffen-Verordnungsblatt, July 13, 1942. This order was numbered 1744 and dated June 26, 1942.

On April 29, 1944, the above pendant was superseded by a golden bar with the appropriate operational number of flights in black (i.e. 200, 300, 400, up to 2000). It was suspended from the Flying Clasp in the same manner as the pendant.[25] The "2000" bar was awarded only once, to Oberst Hans Rudel, in April of 1944. The number "2000" was in diamonds.

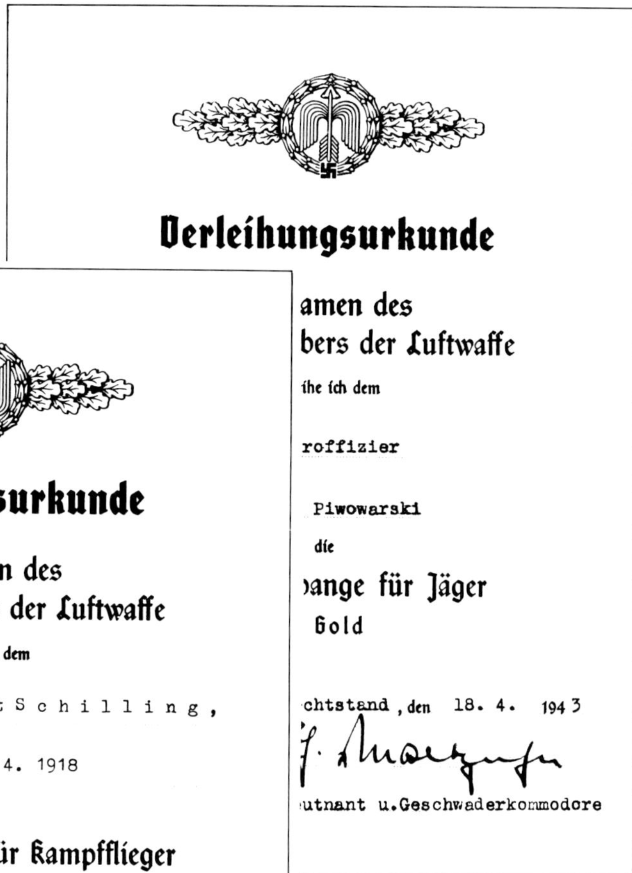

Verleihungsurkunde

…amen des
…bers der Luftwaffe
…ihe ich dem

…roffizier

… Piwowarski

die

…ange für Jäger
… Gold

…chtstand , den 18. 4. 1943

…utnant u.Geschwaderkommodore

Verleihungsurkunde

Im Namen des
Oberbefehlshabers der Luftwaffe

verleihe ich dem

Oberleutnant Helmut S c h i l l i n g ,

geb. am 7. 4. 1918

die

Frontflug-Spange für Kampfflieger
in Silber

Gefechtsstand , den 12. 5. 1941

Major und Kommodore K.G. 30

[25] Luftwaffen-Verordnungsblatt, May 15, 1944. This order was numbered 615 and dated April 29, 1944.

Non-Portable Awards

Goblet of Honor for Outstanding Achievements in the Airwar (Ehrenpokal für besondere Leistung im Luftkrieg)

The Ehrenpokal, instituted by Göring on February 27, 1940, was intended for flying personnel only. It was 20.3 cm high and of 835 silver or Alpaka silver-plate. The obverse carried two fighting eagles in high relief while its reverse displayed an Iron Cross 1st class. A wide pattern of oakleaves and acorns adorned the stem of the Ehrenpokal. Directly below this oakleave pattern was engraved the recipients name, rank and date of award. The legend "Für besondere Leistung im Luftkrieg" appeared in raised letters on the base of the goblet. By further order of Generaloberst Loerzer, the Ehrenpokal was no longer to be awarded to flying personnel after December 10, 1944. In its place the recipient was to receive the Roll of Honor of the Luftwaffe.

It should be noted that this goblet was unofficial and designated for those individuals who had received the Iron Cross, 1st and 2nd class, but who did not merit the Knight's Cross of the Iron Cross.

(obverse)

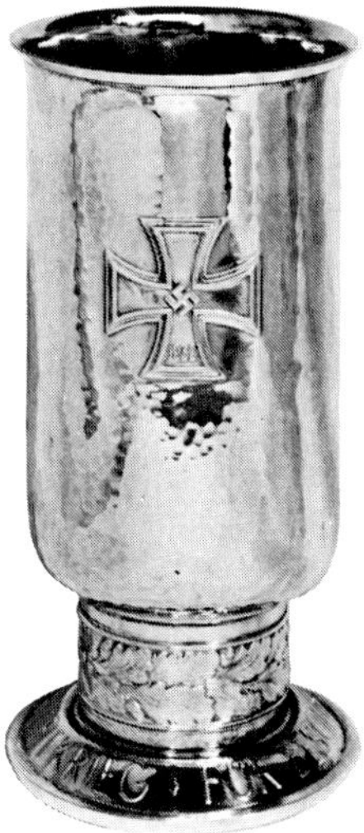

(reverse)

Salver of Honor for Distinguished Achievements in Action (Ehrenschale für hervorragende Kampfleistungen)

The Salver of Honor, instituted by Göring on June 15, 1942, was intended for non-flying personnel in the Luftwaffe (i.e. those in Flak units, paratroopers, members of Luftwaffe field divisions, etc.). The Salver was a finely detailed, shallow plate in silver, 280 mm in diameter. In the center was Göring's Reichsmarschall eagle with the recipient's name, rank, and date of award above and the legend "In Anerkennung hervorragende Kampfleistungen" below. After December 10, 1944, the Salver was no longer awarded. In its place, the recipient was to receive the Roll of Honor Clasp of the Luftwaffe. The Salver was designated for those individuals who had already been awarded the Iron Cross, 1st and 2nd class but did not merit the German Cross in Gold.

NOTE:

Several lesser non-portable awards for achievement were also awarded to Luftwaffe personnel. Detailed coverage of these plaques and medallions may be found in Littlejohn and Dodkins' "Orders, Decorations, Medals and Badges of the Third Reich," pages 189-197.

(The Roll of Honor Clasp for the Luftwaffe is detailed at this time due to its relevance to the Ehrenpokal and Ehrenschale, previously covered.)

Roll Of Honor Clasp (Ehrenblatt-Spange)

By an order dated July 5, 1944, the Roll of Honor Clasp was initiated for Luftwaffe personnel who performed outstanding acts of valor. The clasp consisted of a Luftwaffe eagle with swastika within a wreath of laurel leaves, all of which were in gilt. It was to be worn on the black/white/red ribbon of the 2nd class Iron Cross, which was positioned in the second button hole of the tunic (as for the 1939 bar to the 1914 Iron Cross).[26]

Those awarded this decoration were either already entered or to be entered onto the Roll of Honor of the Luftwaffe. The clasp was also automatically awarded to those individuals who had previously received the Ehrenpokal or Ehrenschale. An order dated December 10, 1944, stated that the Ehrenpokal and Ehrenschale were no longer to be awarded but in their place the Roll of Honor Clasp would be bestowed.[27]

NOTE:

The many decorations and medals permitted to be worn by Luftwaffe personnel, but not exclusively of Luftwaffe origin, have not been included in this publication. Excellent coverage of these numerous pieces may be found, however, in Littlejohn and Dodkins' "Orders, Decorations, Medals and Badges of the Third Reich."

[26] Luftwaffen-Verordnungsblatt, July 17, 1944. The order was numbered 931 and dated July 5, 1944.

[27] Luftwaffen-Verordnungsblatt, December 18, 1944. The order was numbered 1707 and dated December 10, 1944.

The Marksman's Lanyard

The Marksman's Lanyard was awarded to individuals for proficiency with firearms up to and including flak guns. It was worn from the right shoulder to the top button of the Service Tunic or Flying Blouse, with Parade, Walking-Out Dress and Guard uniforms.

Grades 1-4:

were blue-grey silk interwoven with aluminum thread and had a dark silver eagle, wreath and slides. The acorns and antiaircraft shells were in a dark aluminum color (grade 1 - no acorn, grade 2 - 1 acorn, grade 3 - 2 acorns, grade 4 - 3 acorns).

Grades 5-8:

were in aluminum interwoven with blue-grey silk thread and had a matte-aluminum eagle, wreath and slides. The acorns and antiaircraft shells were also aluminum colored (grade 5 - no acorn, grade 6 - 1 acorn, grade 7 - 2 acorns, grade 8 - 3 acorns).

Grades 9-12:

were in blue-grey silk interwoven with gold colored thread and had a gold colored eagle, wreath and slides. The acorns and antiaircraft shells were also gold colored (grade 9 - no acorn, grade 10 - 1 acorn, grade 11 - 2 acorns, grade 12 - 3 acorns).

NOTE:
Marksman's Lanyards for Luftwaffe flak personnel utilized miniature antiaircraft shells in place of acorns.

Marksman's lanyard in aluminum interwoven with blue-grey, silk thread.

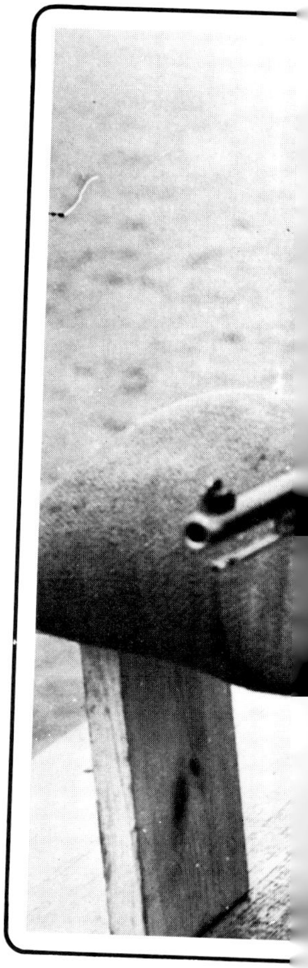

The Unteroffizier at right is wearing the marksman's lanyard, Grade 5.

(Below) A young recruit from Flieger-Ersatzabteilung in Lübeck undergoing proficiency tests with the Mauser rifle

Besitz=
Urkunde
Der Ogfr. Friedr. Marquard

der F.F.Sch.A112 Te.Ko.
hat im Schießjahr 1943/44
in der ersten Schießklasse
für Karab. 98k die
1.STUFE **Schützenschnur** 1.STUFE
— erworben. —

Nellingen, den 1. Okt. 1944

Major u. Komp. Chef

Grade 1 marksman's lanyard award document (for the Mauser 98K)

Luftnachrichten-Helferinnen on break

Civilian and Auxiliary Branches

Civilian Employees in the Luftwaffe

The illustrated insignia for civilian employees and workers in the Luftwaffe was first initiated in matte-silver by an order dated December 13, 1934. It was to be worn on the left lapel of a tunic, jacket or great-coat and on the peaked or forage cap. In 1936, it became available in matte-aluminum and its use became more specific:[1]

(a) worn by civilian employees of the Luftwaffe's fire fighting branch,
(b) as doormen and elevator operators at the RLM building,
(c) as doormen and elevator operators at air district (Luftkreis) office buildings,
(d) or as guards and doormen at specific Luftwaffe posts.

The civilian insignia at left (in aluminum) was introduced in November 1937 and worn exclusively by paid employees of the Reich Aviation Administration on civilian clothing. It was also in gold for individuals with 25 years service. Prior to this date, a Luftwaffe eagle stickpin in gilt metal was worn.

NOTE:
Civilian employee insignia, in stickpin form, could also be worn with civilian clothing . . . on the left lapel for men or as a broach for women.

Civilian Vehicle Drivers

The civilian driver was issued and permitted to wear the following clothing while on duty:

(a) forage cap
(b) tunic
(c) one pair of long pants
(d) one pair of bloused pants
(e) one great-coat in a civilian cut
(f) one pair of black leather leggings
(g) one pair of gloves

The above clothing was in blue-grey and was to be worn only while on duty. The forage cap had the Luftwaffe eagle and cockade.[2] The winged civilian employee insignia in stickpin form (26 mm wide) was worn on the left collar of the tunic.

The tunic was double-breasted with two rows of three buttons, the top button being worn open. All of the above were also issued in mottled-grey for summer wear.

Flying Civilian Employees

This group of specialists was active in training and instruction during the early years of the Luftwaffe. In most cases, they were experienced pilots and wireless operators on loan from Lufthansa. Their distinctive uniform was as follows:[3]

(a) peaked cap in marine-blue

(b) jacket (double-breasted with two rows of two buttons as for Lufthansa), with two side pockets with flaps and one upper left breast pocket

(c) long pants in marine-blue

(d) great-coat (double-breasted with two rows of four buttons and in trenchcoat cut with waist belt), in marine blue

(e) leather great-coat could also be worn (private purchase)

(f) mottled blue shirt }
(g) black tie } as for the Luftwaffe
(h) black shoes }

(i) brown gloves

The system of grading was as follows:

The peaked cap (as for Lufthansa) was in marine-blue with a black cap band. In the middle of the cap band was positioned the winged civilian insignia (14.5 cm wide).

(a) Flugbetriebsleiter (Flight Operations Leader), which included the grades of Stellvertreter, Fl.Kapitäne, as well as Flugzeugführer.......civilian cap insignia in gold

(b) Bordfunker (Wireless Operator) and other flying civilian employees in the Luftwaffe.......civilian cap insignia in aluminum

The cap cords for:

(a) Flight Operations Leaders.......gold interwoven with black

(b) Wireless Operators.......aluminum interwoven with black

(c) other flying civilian employees.......blue

Civilian pilots' badge

[1] Luftwaffen-Verordnungsblatt, February 7, 1938. The order is dated February 3, 1938.

[2] Luftwaffen-Verordnungsblatt, September 2, 1935. The Luftwaffe eagle on the forage cap was later replaced by the national eagle by an order dated January 26, 1938.

[3] Luftwaffen-Verordnungsblatt, April 27, 1936. The order is numbered 488 and dated April 3, 1936.

Uniforms for Doormen and Elevator Operators in the RLM and other Luftwaffe Office Buildings

The following uniform was worn by paid civilian personnel in the capacity of doorman or elevator (lift) operator in buildings housing the offices of:

(a) the Reich Air Ministry (Reichsluftfahrministerium (RLM))

(b) 1. an Air Fleet Command (Luftlottenkommando)

2. a Luftwaffe Commanding Officer (Lw.-Befehlshaber)

3. an Air District Command (Luftgaukommando)

4. a Flying Corps (Fliegerkorps)

The Peaked Cap

The peaked cap for all grades of doormen and elevator operators was of a blue-grey material, with a 4.3 cm wide cap band of either mignonette-green (RLM buildings) or bordeaux-red (other Luftwaffe buildings), a black cap cord and a black leather peak (visor). Above and below the cap band and around the crown of the cap was a 2 mm wide piping in either mignonette-green or bordeaux-red. The insignia for this cap was the national eagle in aluminum on the front portion of the cap and that of the civilian employee in aluminum on the front center of the capband.

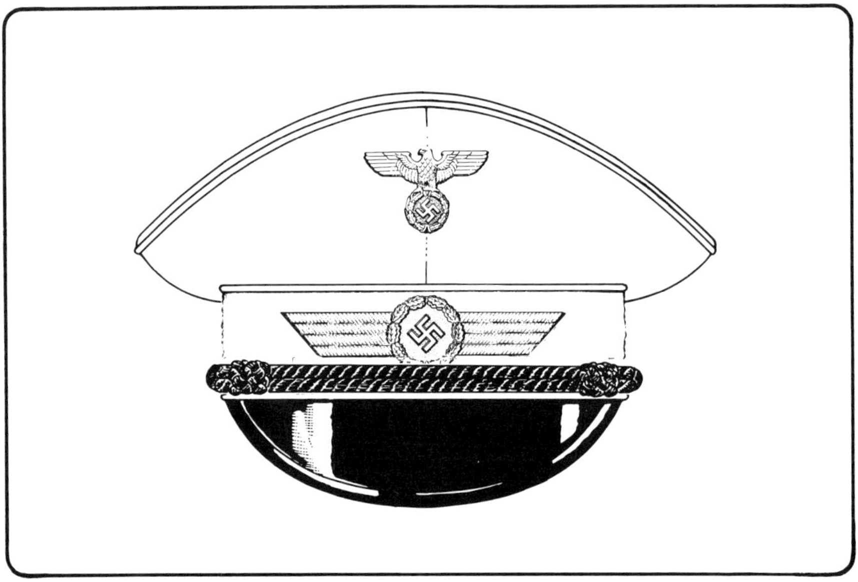

Peaked cap for doormen and elevator operators

Winged insignia for civilian employees in the Luftwaffe

The Uniform

There were three grades of doormen and they were as follows:

(a) Head Doorman (Oberpförtner)

(b) Doorman (Pförtner)

(c) Assistant Doorman (Hilfspförtner)

All three grades plus the one elevator operator grade wore a blue-grey tunic with two rows of three aluminum buttons. The collar was in mignonette-green for those stationed in RLM buildings and in bordeaux-red for the other above mentioned office buildings. Distinctive insignia designating grade was worn in the corners of the mignonette-green or bordeaux-red collars (see illustrations). On the upper left sleeve, was worn a 10 cm wide mignonette-green armband (RLM buildings) with the national eagle in grey thread on a round blue-grey base; and the same armband but in bordeaux-red for other Luftwaffe office buildings (see illustrations). Aluminum Tresse on the top or bottom of the armbands indicated the various grades. Long, blue-grey pants piped in mignonette-green or bordeaux-red were worn with this tunic. A blue shirt, black tie and black shoes were also worn.

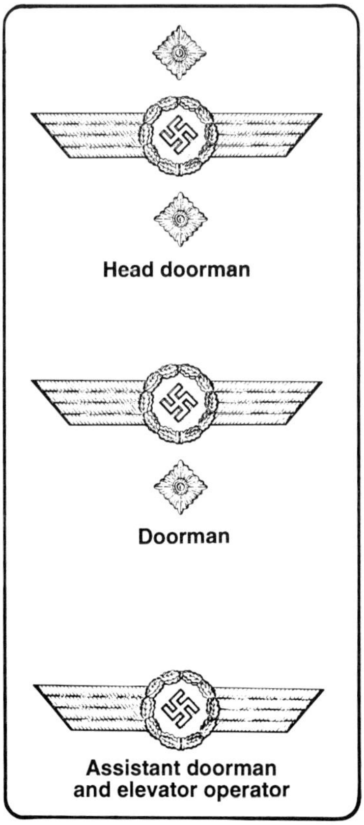

Tunic for doorman rank

Winter Great-Coat

A blue-grey great-coat, of standard Luftwaffe issue, with wool lining was also issued for wear. It had two rows of six aluminum buttons. No shoulder or collar insignia was worn with this great-coat. A raincoat, which was standard Luftwaffe issue, was also issued. The mignongette-green or bordeaux-red armband was worn with both of the above.[4]

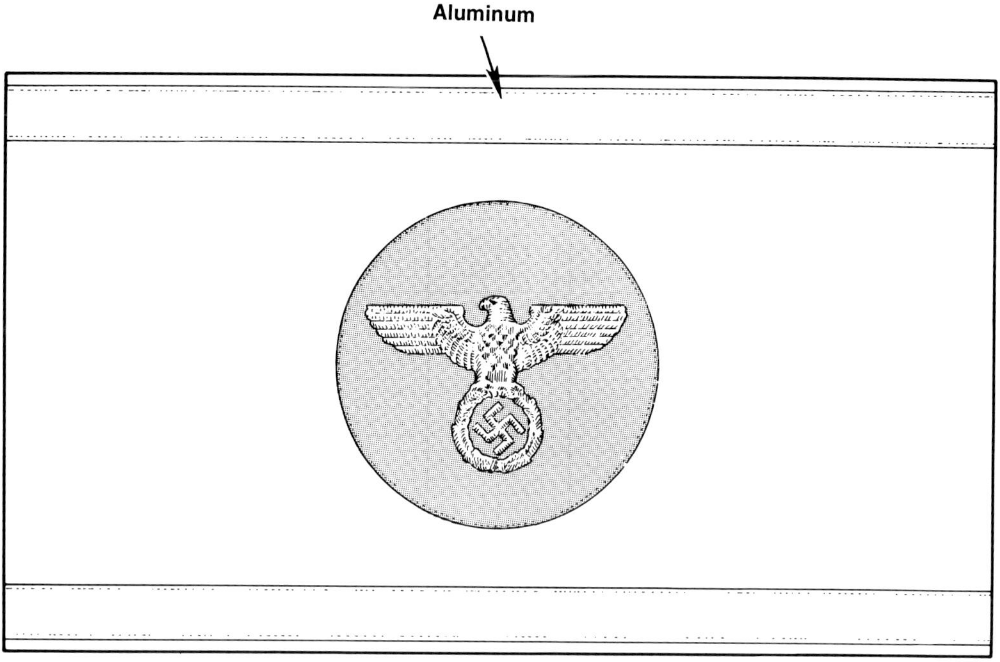

Head doorman's armband (Oberpförtner)

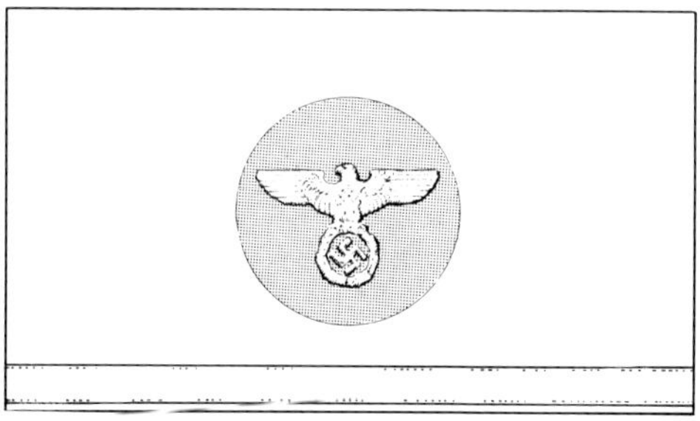

Doorman's armband (Pförtner)

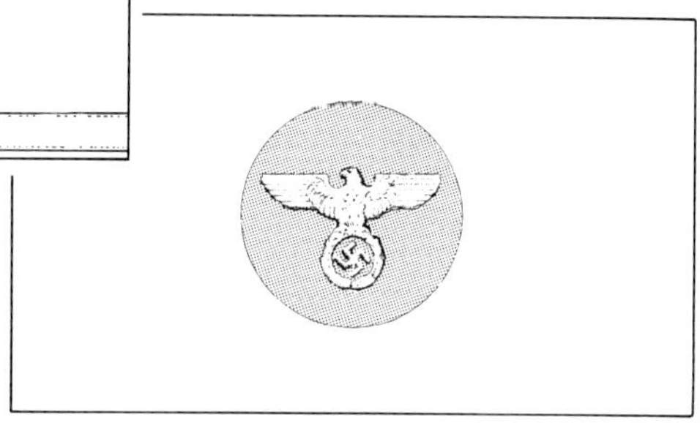

Assistant doorman's and elevator operator's armband (Hilfspförtner u. Fahrstuhlführer)

NOTE:
There was only one grade of elevator operator. Regulations stated that the doorman or elevator operator could wear this uniform only during duty hours.

[4] <u>Luftwaffen-Verordnungsblatt</u>, July 7, 1941. The order is numbered 714 and dated June 24, 1941.

Heimatflak

As the war progressed, Allied bombing of German industrial centers was increased. Therefore, Germany was forced to augment its air defense system with auxiliary units in industrial areas . . . the Heimatflak. The Heimatflak units consisted of plant and office employees who were usually elderly workers or deferred civilians. This new source of manpower was broken into light antiaircraft batteries which ringed their area of employment.

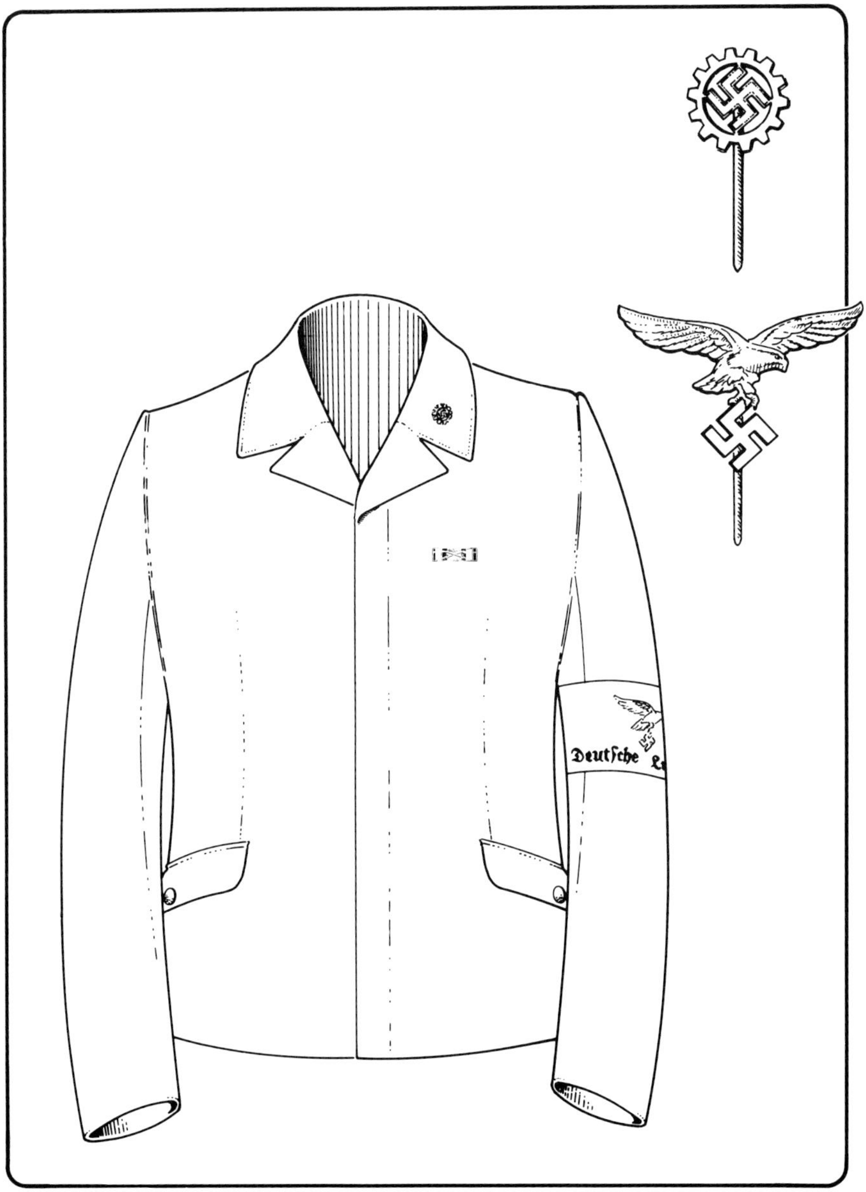

An elderly member of the Heimatflak

Each member of the battery was issued a standard Luftwaffe grey or black, one-piece work uniform (no insignia) or a Flying Blouse (no insignia). They were also issued a Luftwaffe forage cap plus a helmet. On occasion, a "Deutsche Wehrmacht" armband (black on yellow) or a "Deutsche Luftwaffe" armband (blue on white) was seen worn on the left upper sleeve.

The flak soldiers* of the Home and Alarm Flak Artillery also wore insignia denoting their community responsibility. They were as follows:

(a) for working men (a cogged wheel and swastika)

(b) for civilians in the service of the Luftwaffe (a Luftwaffe eagle)

This insignia was worn in stickpin form on the left collar of their work or civilian jacket. The battery crews were trained after work hours by Luftwaffe cadres. Upon completion of a basic training period, the battery commander awarded the civilian individual with the stickpin insignia. Upon separation from the Heimatflakartillerie, however, the insignia was to be returned:[5]

*Flakwehrmänner

[5] Luftwaffen-Verordnungsblatt, March 29, 1943. The order is numbered 602 and dated March 9, 1943.

Flakhelfer*

The Flakhelfer were members of the Hitler Youth who were called upon to assist antiaircraft crews and the Luftschutz during Allied air attacks. They served as messengers, weather observers, signalers and in some cases as gun crew personnel in antiaircraft towers.

The Hitler Youth who remained in school or worked in the defense industries were essential to the Reich's air raid protection measures. In the industrial areas, these boys were assembled into Hitler Youth fire brigades and participated in night air raid alarms. In early 1943, an agreement between the Reich's Youth Leader, Artur Axmann and the SS-Reichsführer, Heinrich Himmler, permitted police authorities to call on all Hitler Youth boys fifteen years or older for participation in fire fighting units.

Luftwaffenhelfer*

The Luftwaffenhelfer, also an HJ (Hitler-Jugend) organization, was organized to accelerate the transformation of HJ members into Luftwaffe personnel. To accomplish this, Luftwaffe training centers were established to complete the training the HJ members had already received in the HJ Motor Corps and the NSFK schools. Here, under Luftwaffe guidance, the Hitler Youth members received instruction in military discipline, ground antiaircraft defense and maintenance of ground defense equipment. Those members who showed exceptional ability were sent to special schools for pilot and bombardier training.

***In reality the same organization, but with different specialties.**

A Luftwaffenhelfer and member of a Berlin flak battalion, shortly after being awarded the Luftwaffe flak badge for outstanding service against "terrorist bombers."

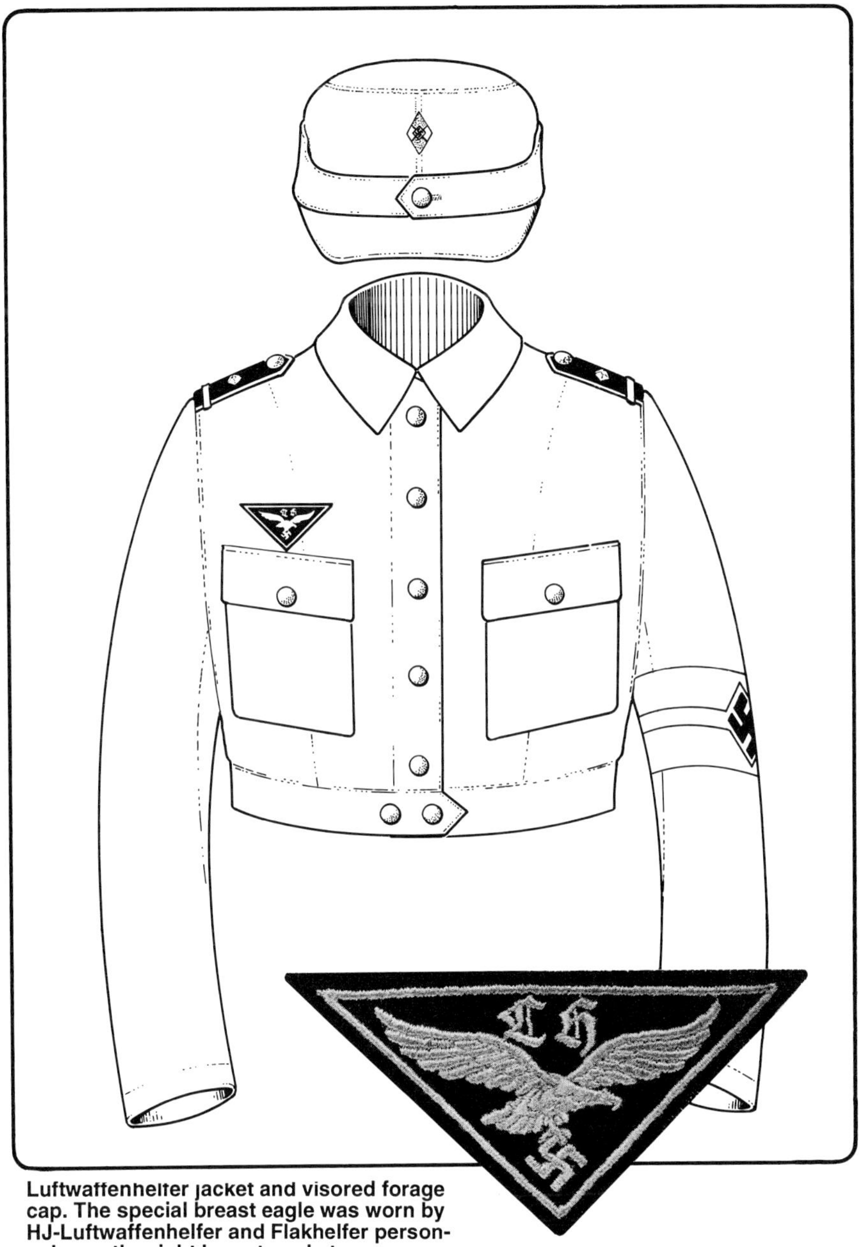

Luftwaffenhelfer jacket and visored forage cap. The special breast eagle was worn by HJ-Luftwaffenhelfer and Flakhelfer personnel over the right breast pocket.

The Luftwaffenhelfer uniform consisted of a waist-length jacket with two breast pockets, long pants and a visored forage cap, all in blue-grey. The jacket's insignia consisted of light-blue piped Hitler Youth shoulder straps, the Hitler Youth armband, the triangular district patch on the upper left sleeve (ordered removed later in the war) and a special "Luftwaffenhelfer" breast eagle (light blue on black). The visored forage cap carried only the Hitler Youth diamond insignia.

Those Luftwaffenhelfer members assigned to transport duty, such as accompanying supply trucks or trains or food dispensing vehicles, were not permitted to wear the "Luftwaffenhelfer" breast eagle. They could, however, wear their district patch again while on transport duty.[6]

Luftwaffenhelferinnenschaft*

It was inevitable that women would be mobilized within the Reich as the demands of war took an increasing toll of German manpower. Starting in early 1940, women were recruited into the Nachrichtenhelferinnen des Heeres and the Luftwaffenhelferinnen, thus releasing thousands of male clerical personnel for combat duty on the much-pressed fronts (i.e. six soldiers were replaced by nine Helferinnen). Positions for recruits were available as clerks, telephone, teletype and radio operators, and intelligence service assistants (all trained by the Luftnachrichtentruppe).

Women between the ages of seventeen and forty-five, familiar with office procedures, were considered for domestic and foreign assignments. Luftwaffenhelferinnen volunteers were stationed as key office personnel in the occupied countries as well as in Germany. Late in the war, with increased Allied bombing, the Luftwaffenhelferinnen were often relieved of their office duties to assist in air-raid protection. To fill the depleted Luftwaffe ranks, sections of the female auxiliaries were also trained as search-light crews and flak-crew assistants (in 1944 also medical).

Luftnachrichten-Helferinnen

The Ln-Helferinnen, a part of the Luftwaffenhelferinnen, consisted of the Flugmeldedienst (Air Warning Service) and the Fernsprech-und Fernschreibbetriebsdienst der Luftwaffe (Telephone and Teletype Department of the Luftwaffe). The Flugmeldedienst was given the responsibility of physical observation of the sky, identification and direction of aircraft, the reporting of these sightings, and the use of searchlights, binoculars, and hearing apparatus. Their uniform was as follows:[7]

1 - forage cap, blue-grey
1 - winter great-coat, double breasted with two rows of four buttons, blue-grey
1 - rain cape
1 - suit which consisted of:
1 - straight skirt with pleat, blue-grey
1 - jacket, three front buttons, two side pockets with flaps and one inner left pocket, blue-grey
1 - work smock

***1940: Weibliches Fernsprechbetriebspersonal, after 26.2.1941: Ln-Helferinnenschaft with Flugmeldediensthelferinnen and Luftnachrichtenhelferinnen.**

[6] Luftwaffen-Verordnungsblatt, June 19, 1944. The order is numbered 819 and dated June 5, 1944.

3 - service blouses with collar, light blue
1 - service blouse with collar, white
2 - black ties
1 - purse with shoulder strap
2 - pairs of black shoes

Rank Insignia

1. Their rank insignia to June 1940 was standard Luftwaffe issue.
2. From June 1940 to July 1941, the following ranking system was utilized:[8]

(No insignia)

Anwärterin (after August 1940-Flugmeldehelferin)

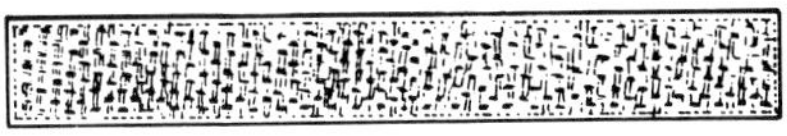

Helferin (after August 1940-Flugmeldehelferin)

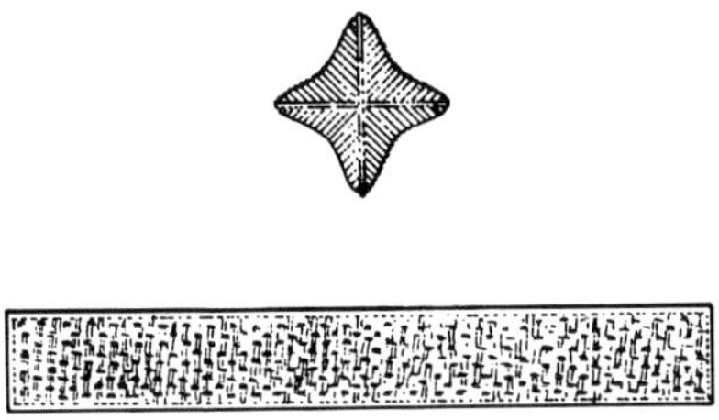

Aufsicht (after August 1940-Aufsichtshelferin)

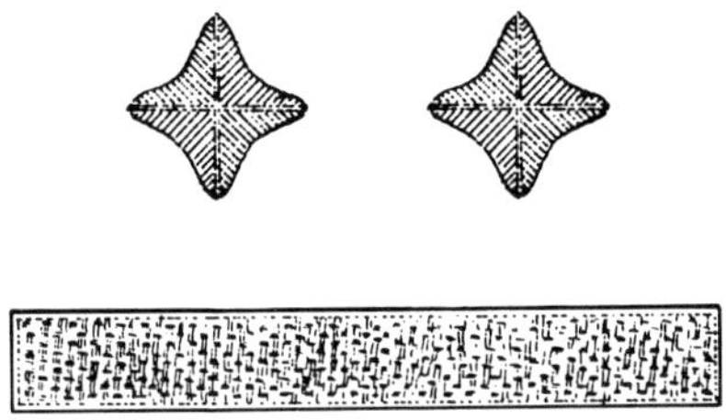

Betriebsgruppenführerin (after August 1940 Betriebsgruppenunterführerin)

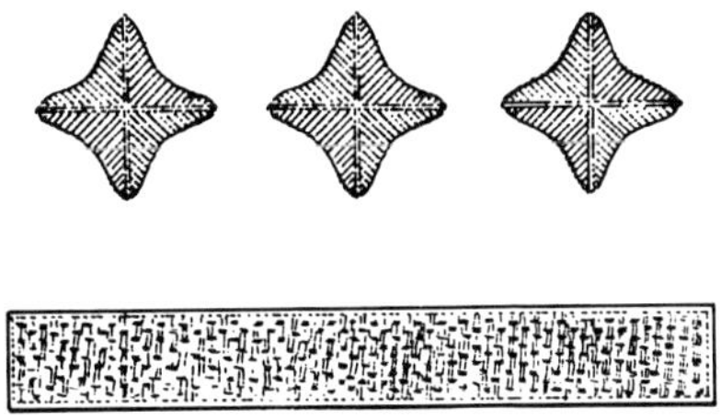

Hauptgruppenführerin (after August 1940-Betriebs-Gruppenführerin und Heimleiterin)

Note positioning of Hauptgruppenführerin rank insignia at right.

[7] Luftwaffen-Verordnungsblatt, August 11, 1941. The order is numbered 896 and dated July 28, 1941.

[8] Luftwaffen-Verordnungsblatt, July 1, 1940. The order is numbered 794 and dated June 11, 1940.

3. On July 28, 1941, a further change in the ranking system was initiated by the Ln-Helferinnen and utilized for the duration of the war.[9]

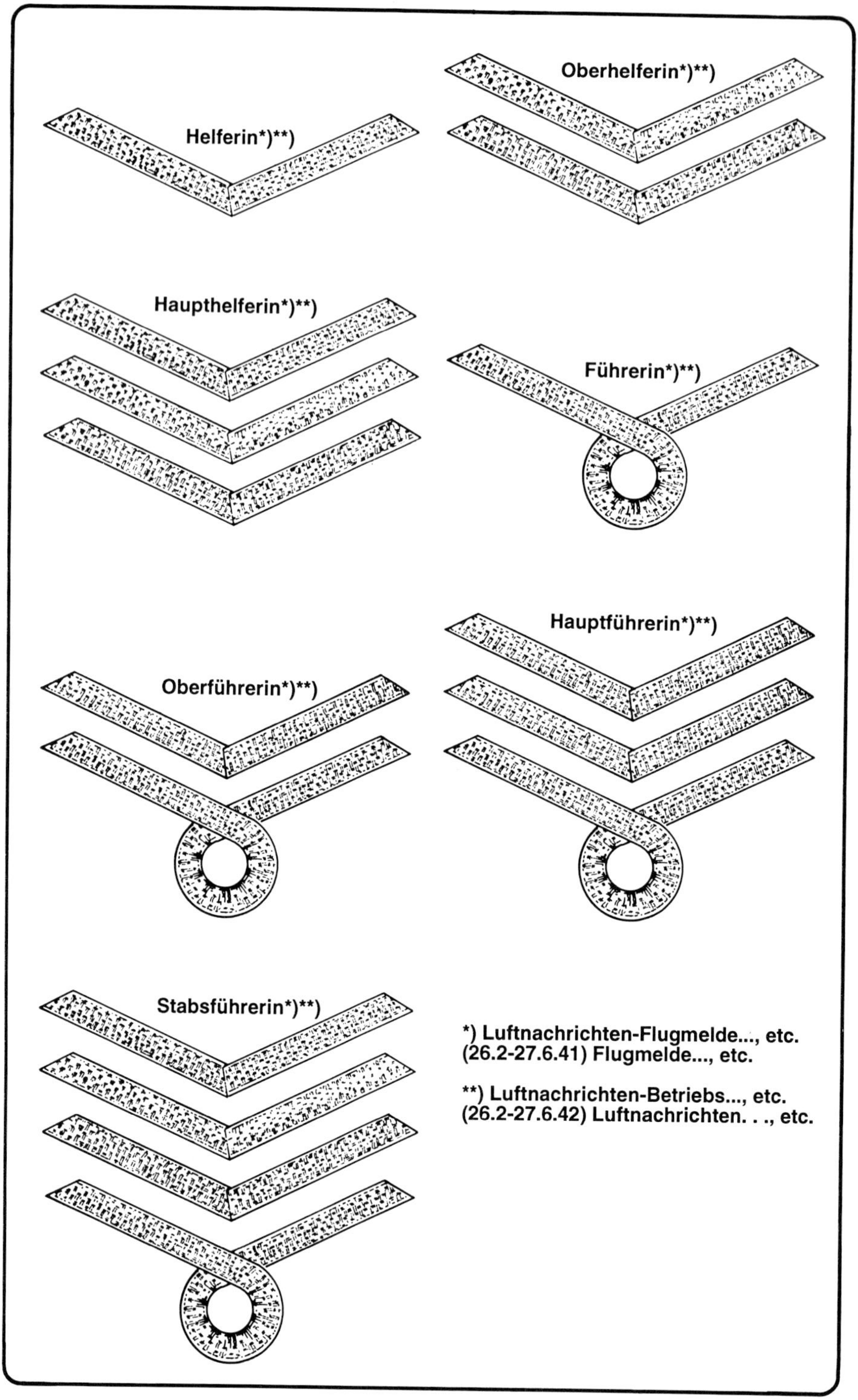

NOTES:
1. The chevron was worn 1 cm below the specialty patch.
2. Führerinnen wore an aluminum embroidered star on both collars of the suit jacket and great-coat. The collar was also piped with a 3 mm aluminum cord after 29.10.43 (also forage cap).

Uniform for a Führerin in charge of a Ln-H-Betriebszug

Members of the Ln-Helferinnenschaft wore with their service clothing:[10]

1. Luftwaffe eagle in matte-grey, machine-embroidered thread (in aluminum for Führerinnen)

(a) on the forage cap
(b) on the right breast of the jacket } ordered 26.11.40

[9] Luftwaffen-Verordnungsblatt, August 11, 1941. The order is numbered 896 and dated July 28, 1941.

[10] Ibid.

2. Luftwaffe eagle (aluminum) in stickpin form

(a) for the black tie

3. golden-brown piping, 3 mm wide, around the turn-up of the forage cap

4. specialty patches, worn on the left sleeve of the jacket and great-coat 12 cm below the shoulder seam

5. (a) Führerin up to Stabsführerin in charge of a Flum-H.-Einsatzgruppe of a Ln-H.Betriebszug (1 Ln-Führerin and 2 - 4 Betriebsgruppen*) wore a 0.5 cm aluminum stripe on both sleeves of the suit jacket, great-coat or smock (12 cm from the bottom of the sleeve).

*1 Ln-H.-Betriebsgruppe consisted of 1 Haupthelferin and up to 8 Helferinnen.

(b) Führerin up to Stabsführerin in charge of a Flum-H.-Kameradschaft or a Ln-H.-Betriebskameradschaft (1 Ln-Oberführerin and 2 - 4 Betriebszüge) wore two 0.5 cm aluminum stripes on both sleeves of the suit jacket, great-coat or smock.

NOTE:
If appointed to the above positions, Ober- and Haupthelferinnen also wore the sleeve stripes.

Luftschutz-Warndienst

The LS-Warndienst was a uniformed civilian organization whose job was to take the reports from the Flugmeldedienst, analyse them and sound the alarm at the Luftschutz Central Warning Office. This in turn triggered the civilian warning systems. They were also charged with sounding the "all clear." Female personnel of the Luftschutz-Warndienst were issued the same clothing as the Ln-Helferinnen. Their insignia and ranking system was different, however, and was as follows:[11]

1. The Luftschutz emblem in matte-grey, machine-embroidered thread (in aluminum for Führerinnen)

(a) on the forage cap

(b) on the right breast of the jacket

2. The Luftschutz emblem in stickpin form (white metal)

(a) for the black tie

3. The LS-Warndienst insignia on the upper left sleeve of the jacket

4. green piping, 3 mm wide, around the turn-up of the forage cap

5. rank insignia worn 10 cm from the bottom of the left sleeve (jacket and great-coat) (after 3.2.42, rank insignia was as for Luftnachrichten-Helferinnenschaft)

The Luftschutz emblem for the jacket and forage cap

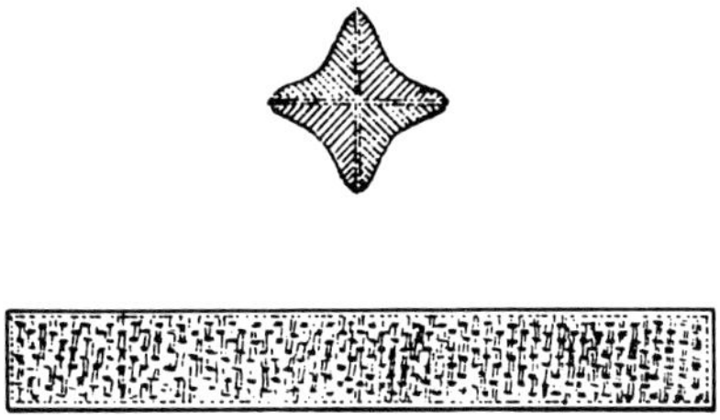

LS-Warndiensthelferin

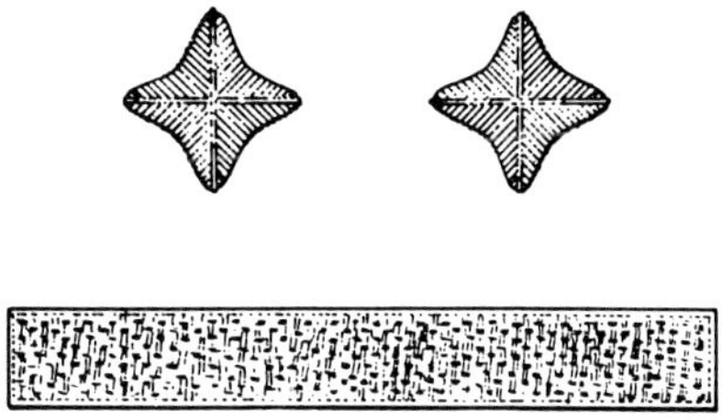

LS-Warndienstoberhelferin

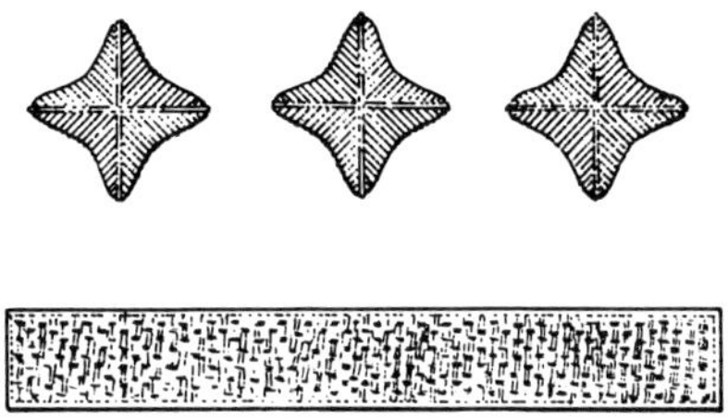

LS-Warndiensthaupthelferin

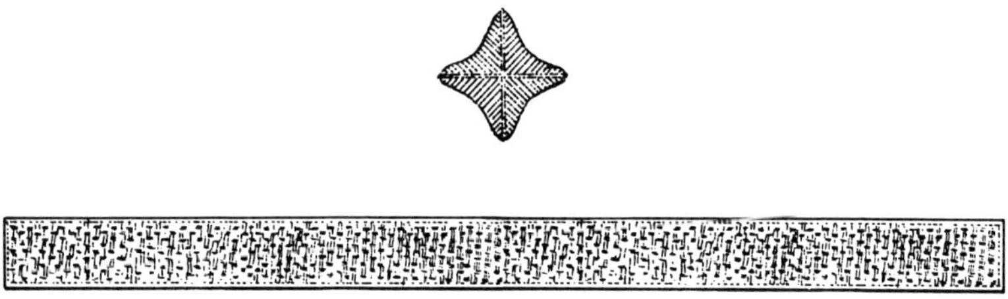

LS-Warndienstführerin
(1 cm wide tresse goes around the sleeve)

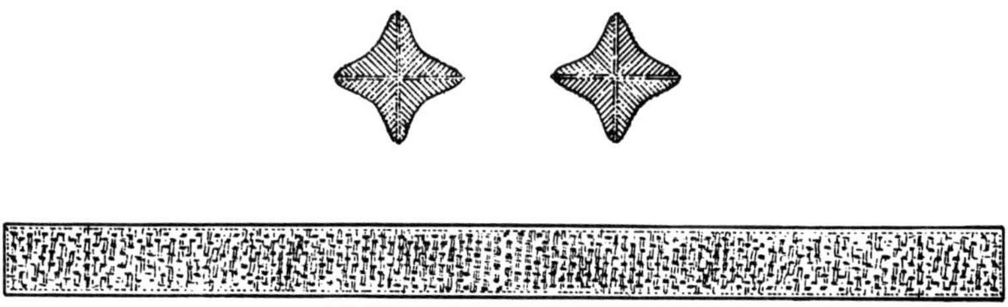

LS-Warndienstoberführerin
(1 cm wide tresse goes around the sleeve)

Luftwaffenhelferinnen

Those Luftwaffenhelferinnen personnel not belonging to the previously mentioned sub-units underwent an insignia change on March 17, 1944, which was as follows:

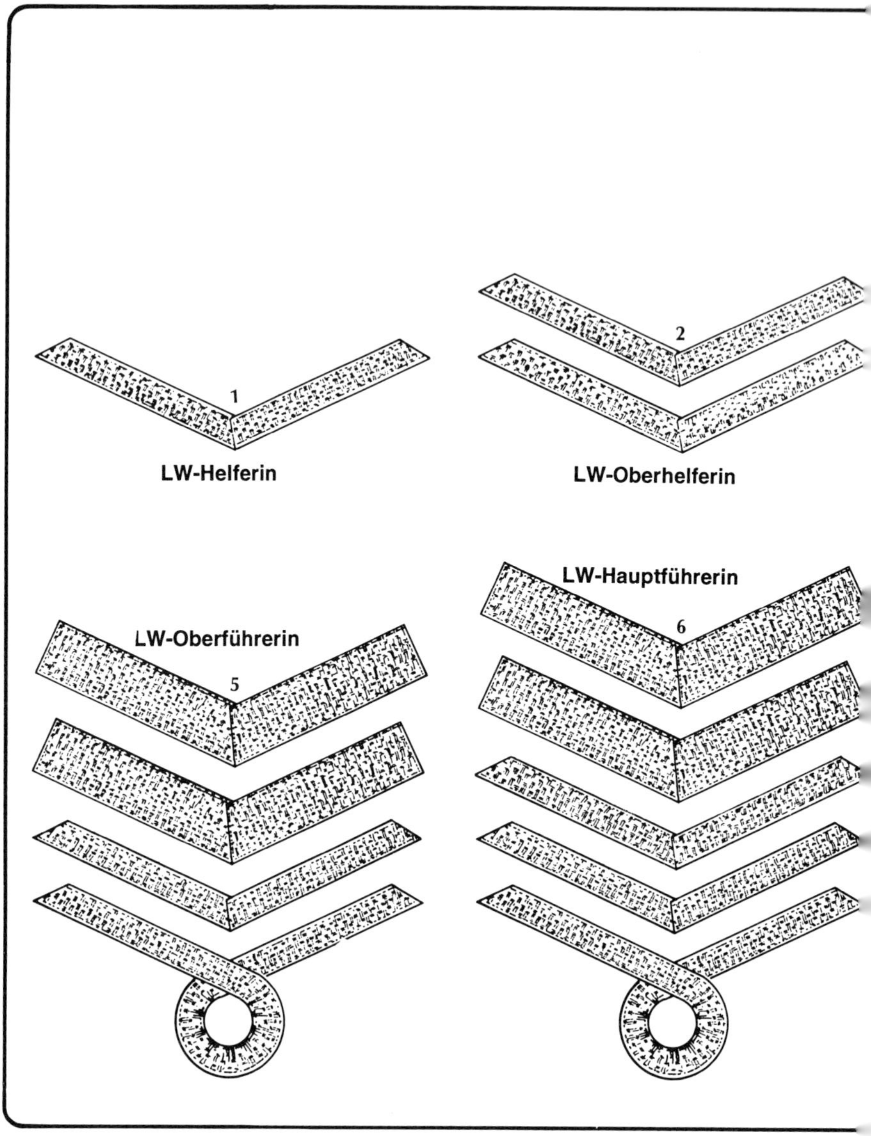

Flakwaffenhelferinnenkorps (after 16.10.43)

The Flakwaffenhelferinnen were not members of the armed forces but were, however, governed by military regulations. Personnel usually consisted of volunteers from the Luftwaffenhelferinnen but if emergencies warranted, the local department of labor could assign additional individuals to this unit. Flakwaffenhelferinnen were always stationed

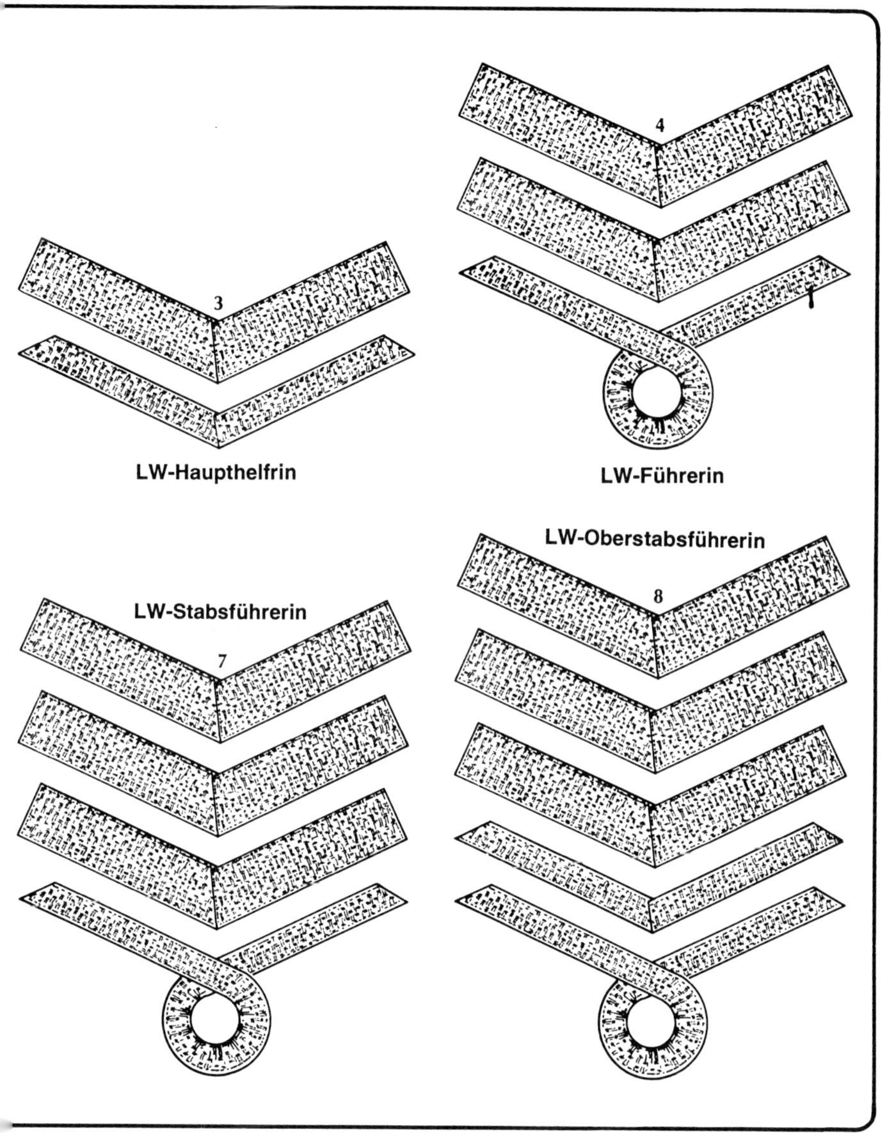

within the Reich, usually near fixed battery sites and operated searchlights, listening devices and German radar apparatus. The Flakwaffenhelferinnen uniform consisted of a jacket with hidden buttons (as on the flying blouse) with belt and two side pockets, pants and a single button M43 forage cap. A special Flakwaffenhelferinnen insignia was worn on the upper right sleeve.

Flakwaffenhelferin arm patch

Flakwaffenhelferinnen Rank Insignia

(see previous page for number code)
Flakw.-Helferin - (as #1)
Flakw.-Oberhelferin - (as #2)
Flakw.-Truppführerin - (as #3)
Flakw.-Obertruppführerin - (as #3 but with two thinner chevrons)
Flakw.-Führerin - (as #4)
Flakw.-Oberführerin - (as #5)
Flakw.-Hauptführerin - (as #6)
Flakw.-Stabsführerin - (as #7)
Flakw.-Oberstabsführerin - (as #8)

1. The above rank insignia was worn on the upper left sleeve.
2. Ranks of Führerin to Oberstabsführerin had an aluminum star on both collars (also piped with a 3 mm aluminum cord).

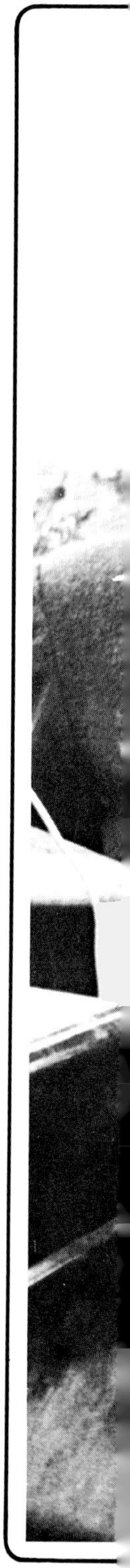

Flakwaffenhelferinnen on duty. The small black patch on the Helferin's collar in background indicates she is in mourning for a friend or relative.

In 1944, the "Ostkampfhelferinnen der Luftwaffe" was formed and consisted mainly of eastern European refugees. On November 13, 1944, however, the title was changed to "Luftwaffenkampfhelferinnen" due to the large influx of refugees, not only from the east, but from all European countries.

Also in early 1945, all wives of Army, Navy and Luftwaffe personnel were formed into the "Wehrmachthelferinnenkorps." The Corps was broken into three main groups: Stabshelferinnen, Nachrichtenhelferinnen and Truppenhelferinnen (the latter were assigned to transport duty, Flak duty, etc.). Although ordered on February 1, 1945, the building of the "Helferinnenkorps der Luftwaffe" (Wehrmachthelferinnen (Luftwaffe)) did not proceed until March 8, 1945. The uniforms were as previously described but with the following rank insignia on the lower left sleeve.

(a) Helferinnen
  1. Helferinnen (=Grenadier)
  2. Oberhelferin (=Gefreiter)
  3. Haupthelferin (=Obergefreiter)

(b) Unterführerinnen
  4. Truppführerin (=Unteroffizier)
  5. Obertruppführerin (=Feldwebel)

(c) Führerin
  6. Dienstführerin (=Leutnant)
  7. Oberdienstführerin (=Oberleutnant)
  8. Hauptdienstführerin (=Hauptmann)
  9. Stabsführerin (=Major)
  10. Oberstabsführerin (=Oberstleutnant)

(no insignia)

2

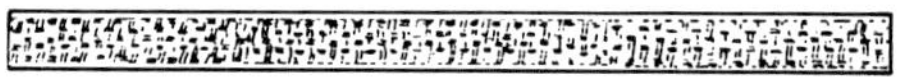

3

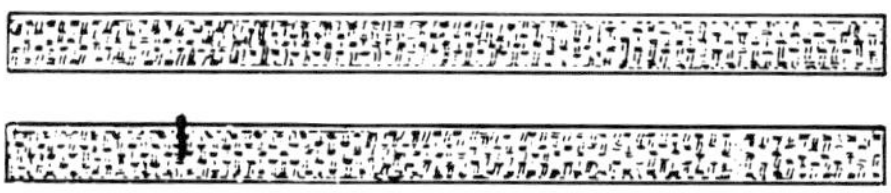

4

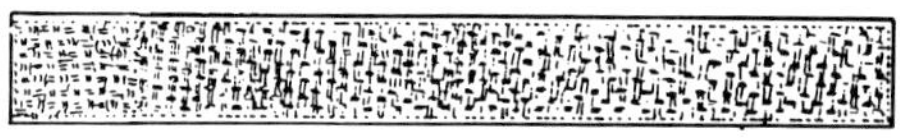

5

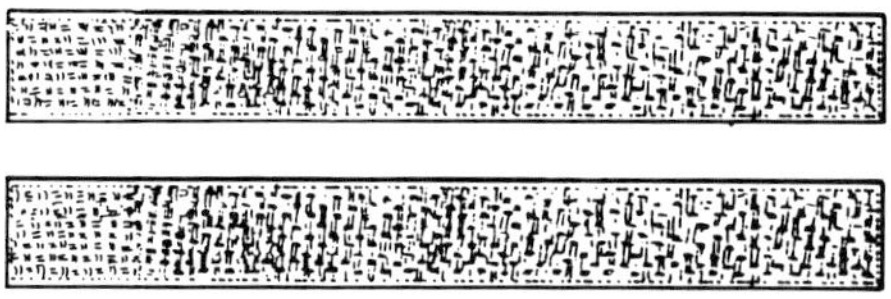

6

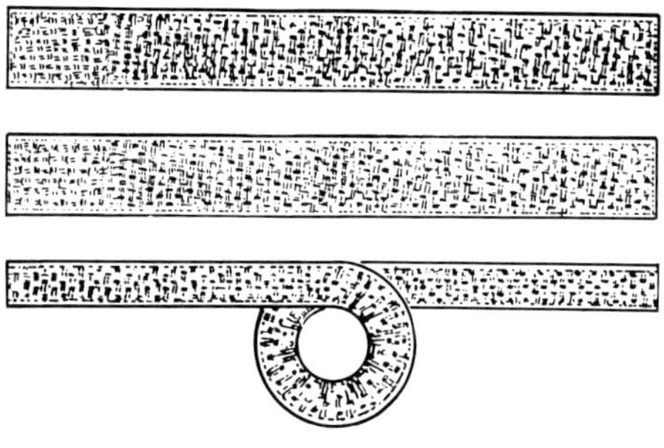

7

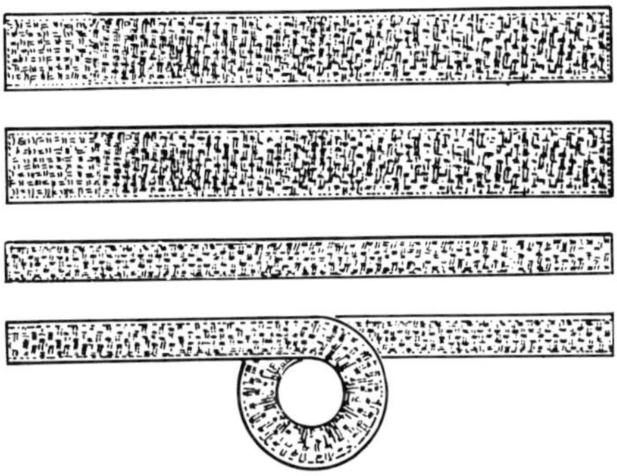

8

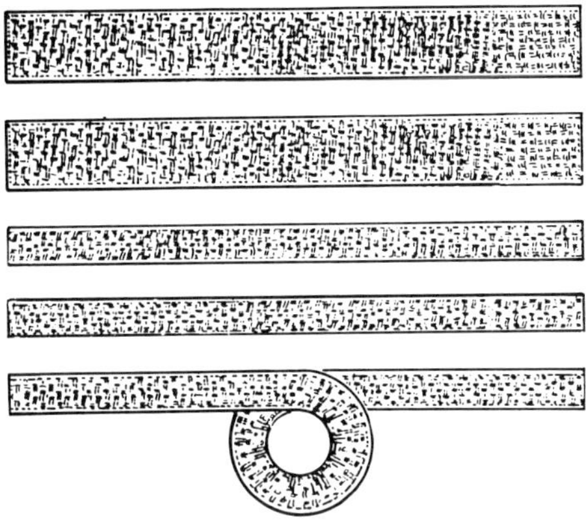

9

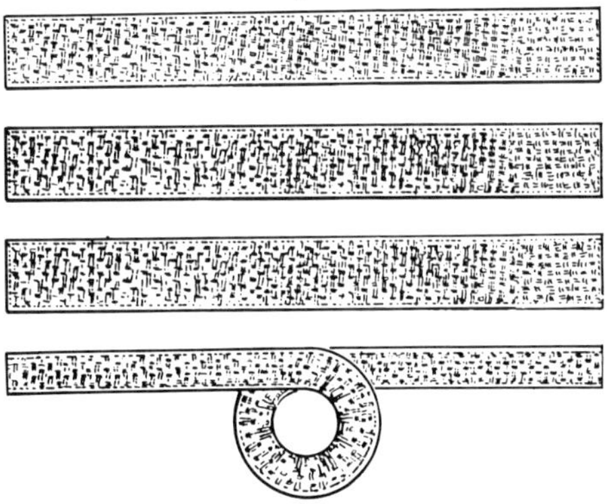

10

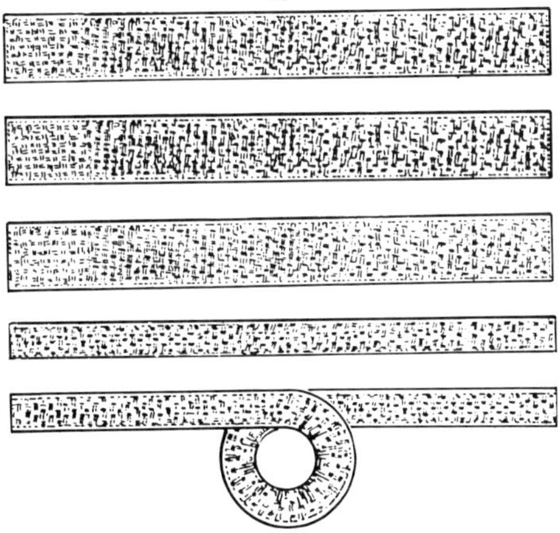

The order dated March 8, 1945, also called for the creation of two new appointed positions . . . Generalführerin and Vertreterin der Generalführerin. The rank insignia for these two high positions are not known.

NOTE:

The female RAD (Kriegshilfsdienstmaiden) were also active in home defense. They were originally with the Flugmeldedienst (1943) and later assigned to radar sites. In the spring of 1944, their jobs were expanded to help man Flak batteries. And finally in January 1945, they took over all searchlight batteries. The female RAD uniform was worn.

Generalfeldmarschall von Blomberg presenting the Luftwaffe standard to Kampfgruppe "Hindenburg" on November 11, 1936.

Standards, Flags and Pennants of the Luftwaffe

The Luftwaffe Standard (Fahne)

By a decree dated March 16, 1936, (the first anniversary of the rebirth of the German Wehrmacht), Hitler ordered that new troop standards shall be presented to units of the Wehrmacht. This order was put into effect on May 11, 1936. From this date, new Luftwaffe regiments were presented their newly designed standards while Regiment "General Göring" retained its original standard (initially for the Prussian Schutzpolizei).

The Luftwaffe standard is 126 cm x 126 cm and is in three colors (Waffenfarben): Yellow - Flying troops (Fliegertruppe), Red - Anti-aircraft artillery (Flakartillerie) and brown - Signals troops (Luftnachrichtentruppe)

Right side: In the center of the standard, on a white circular field, is a Luftwaffe eagle on a laurel wreath in aluminum wire. Four black and white rays extend from the laurel wreath to the four corners of the standard, where black swastikas bordered in aluminum wire are positioned.

Left side: The basic design is as the right side, but with the exception of an aluminum edged, black iron cross, surrounded by an oakleave wreath in aluminum, in the center. The standard is edged on three sides with an aluminum colored fringe. The fourth side is attached to the pole with aluminum colored braid and tacks.

Accessories

The pole, on which the standard is affixed, is of oak, in a black finish and has an aluminum colored Luftwaffe eagle pole top. Below the attached standard is a silver ring with an inscription of the standard's dedication date and the unit designation. On the lower end of the pole is an aluminum cap.

Two streamers (1.50 m long and 36 mm wide) are also part of the standard's accessories...they are aluminum colored and edged in black, aluminum and red. The tassels (approximately 17 cm long) are the same color configuration.

A bandoleer of Morocco leather, to support the standard, is worn from the left shoulder (over the shoulder strap) with its boot of chocolate-colored Morocco leather hanging at the right hip. The bandoleer is edged with a 4 cm wide aluminum braid with the center section in its appropriate Waffenfarbe.

Standard-bearers (Fahnenträger) wore a special insignia on the upper right arm of the tunic and flight blouse (crossed standards in appropriate Waffenfarbe and aluminum wire embroidery). They also wore a distinctive gorget (Ringkragen) while on duty as a standard-bearer.

Note silver dedication ring on pole plus early bandoleer with its foul weather cover worn by the Feldwebel or Wachtmeister.

Note standard-bearer accessories being worn.

Gorget for Fliegertruppe, Fallschirmtruppe, Flakartillerie and Luftnachrichtentruppe.

Luftwaffe standard (Flakartillerie)
(right side)

NOTE:
OKW (Hitler) order dated 28.8.44 and Luftwaffe (OKL) order dated 4.12.44 stated that all flags and standards (Fahnen und Standarten) are to be turned in where they will be displayed at museums. From this time, the Reichskriegsflagge was to be the color for battalions, etc.

Standard bearers' arm insignia (Flakartillerie)

Luftwaffe standard (Fliegertruppe or Fallschirmtruppe)
(right side)

Luftwaffe standard (Luftnachrichtentruppe)
(left side)

NOTE:

Fallschirjäger battalions carried the same standard as for the Fliegertruppe as their branch color (Waffenfarbe) was golden yellow. The only exception was Fallschirmjäger-Regiment 1 which carried three different colors: I. Bataillon (formerly IV. (Fallsch.-Schützen) Btl. Regt. “General Göring”) carried the Regiment “General Göring” standard without the coat of arms below the laurel wreath; II.Bataillon (formerly Fallschirm-Infanterie-Bataillon of the Army) carried a white dove-tailed standard for motorized units of the Army; and III.Bataillon carried the Luftwaffe standard in golden yellow.

Members of IV.Bataillon, Regiment “General Göring” in the Sudetenland (October 1, 1938).

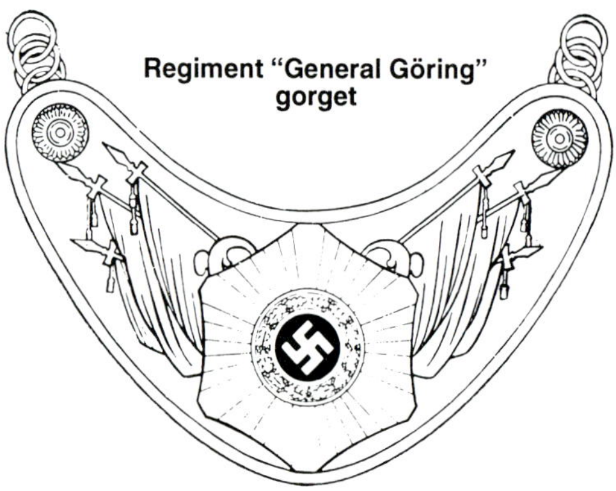

Regiment “General Göring” gorget

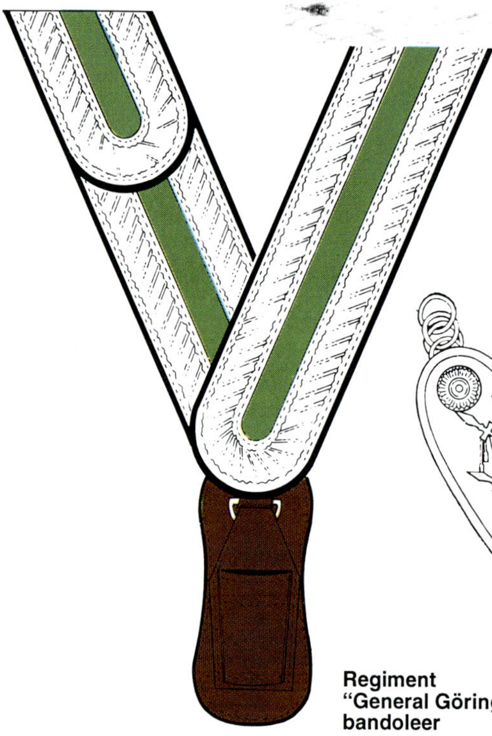

Regiment “General Göring” bandoleer

I.Btl., Regiment "General Göring" (other battalions without "Southern Cross" coat of arms under laurel wreath)

Regiment "General Göring" standard bearers' arm insignia

Göring's Personal Standards and Flags

Kommandoflagge des Reichsministers der Luftfahrt (from 5.5.1933-1935)

NOTES:

1. All measurements of command, staff and personal flags are for vehicles. Double measurements for battle, staff or HQ flags.

2. Stabsflagge: for commanding officers of battalions and regiments.
Kommandoflagge: for commanding officers of brigades and higher.
Dienstflagge: for the chief of an office.
Persönlicheflagge: according to person, rank or Rangflagge.
Wimpel (car pennant): for personal use only.

Kommandoflagge des Reichsministers der Luftfahrt und Oberbefehlshabers der Luftwaffe (after 1935)
(right side)

(left side)

NOTE: by an order dated April 28, 1938, crossed blue batons were now positioned above the Pour le Mérite on the right side.

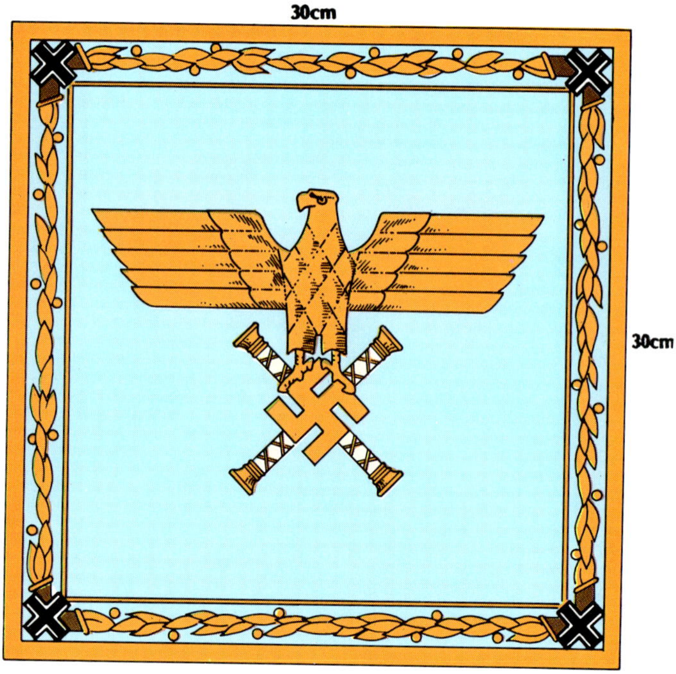

Standarte des Reichsmarschalls des Grossdeutschen Reiches (1st pattern) (left side)

(right side)

NOTE: used from late 1940 to February 17, 1941.

Standarte des Reichsmarschalls des Grossdeutschen Reiches (2nd pattern) (left side)

(right side)

NOTE: order dated February 17, 1941.

Reichsmarschall Göring visiting a Luftwaffe unit in May 1942.

Reichsmarschall Göring at the English Channel during the Battle of Britain.

Rangflagge eines Generalfeldmarschalls der Luftwaffe

(left side)

NOTE: This personal flag was used by all field-marshals of the Luftwaffe. When doing so, a command flag was required to be present (i.e., Generalfeldmarschall Milch would display the above flag plus that of the Generalluftzeugmeister).

30cm

30cm

Kommandoflagge des Staatssekretärs der Luftfahrt (title initiated February 22, 1933 and discontinued January 7, 1945) and Generalinspekteurs der Luftwaffe.

Staatssekretär der Luftfahrt und Generalinspekteur der Luftwaffe Milch (later replaced Ernst Udet as Generalluftzeugmeister upon his death).

Kommandoflagge des Chefs des General Stabes der Luftwaffe (title held by Hans Jeschonnek until his suicide on August 18, 1943) (discontinued on July 13, 1940)

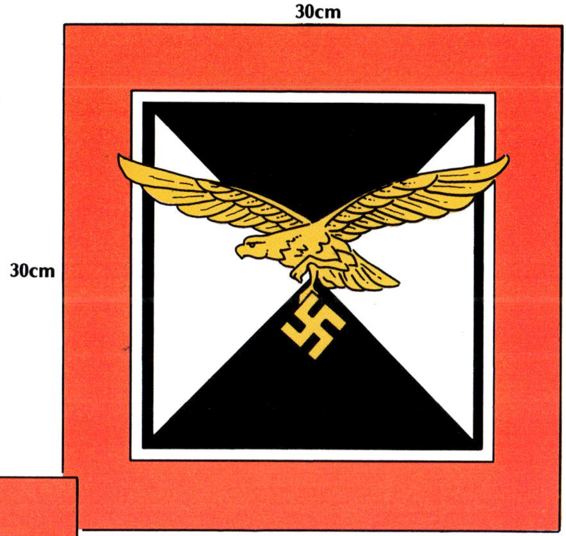

Dienstflagge des Generalluftzeugmeisters (initiated July 13, 1940 and discontinued June 20, 1944)

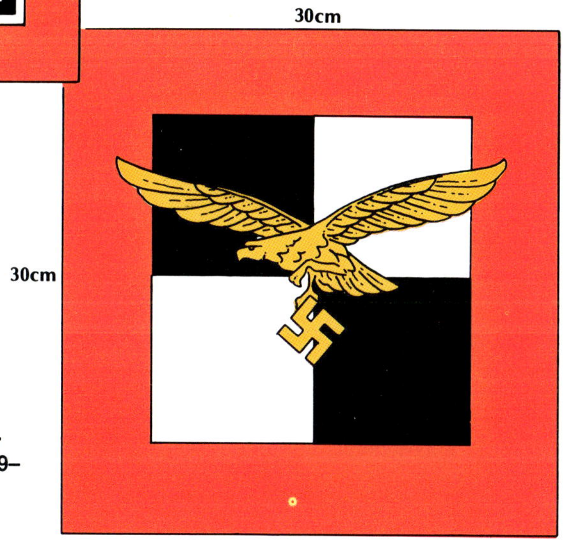

Kommandoflagge des Chefs Einer Luftlotte (initiated February 1, 1939– also Fallschirm-Armeeoberkommando 1944/45).

(Before July 13, 1940) Commanding General and Commanding Officer of a Luftkreis (1.4.1935-4.2.1938) or Luftwaffengruppen Kommando (4.2.38-1.2.39)

After 13.7.1940:

(a) Commanding General of a Flying, Flak or Fallsch.-Corps

(b) Commanding General and Commanding Officer of a Luftgau

(c) Luftwaffe General attached to the Army High Command

(d) Luftwaffe General attached to the Navy High Command

Additionally after 29.4.1943:

(e) Commanding General of a Luftwaffe Field-Corps or Fallschirm Panzerkorps

Division-level Commanders (or higher commanders of demonstration units, etc.)

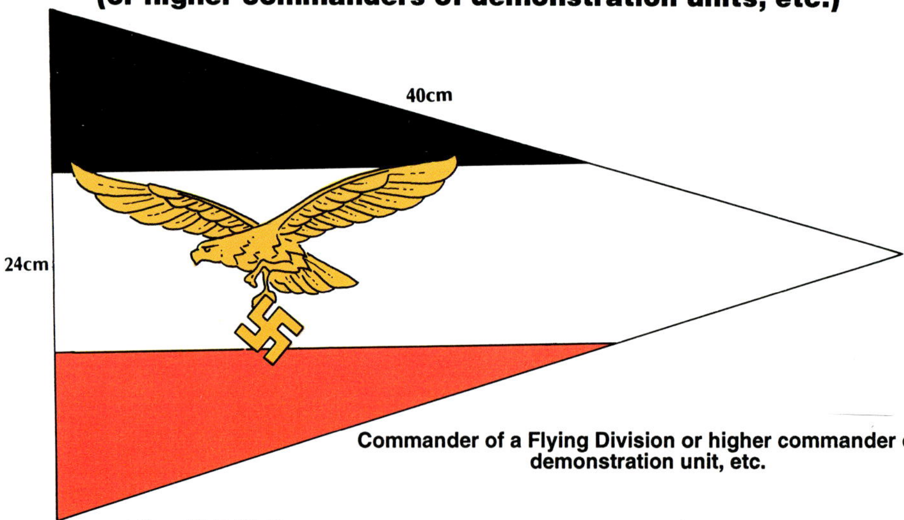

Commander of a Flying Division or higher commander of a demonstration unit, etc.

After 13.7.1940:

(a) Commander of a Flying Division

(b) Higher Signals Leader

(c) Leader of the Sea Air Force

(d) Commanding Officer of the Fortress Anti-Aircraft
(e) Commanding Officer of an Air Defense Area
(f) Luftwaffe Commander attached to an Army Group
(g) Commander of a Fallschirmjäger-Division

Additionally after 29.4.1943:
(h) Commander of a Luftwaffe Field Division
(i) Commander of Division "Hermann Göring"

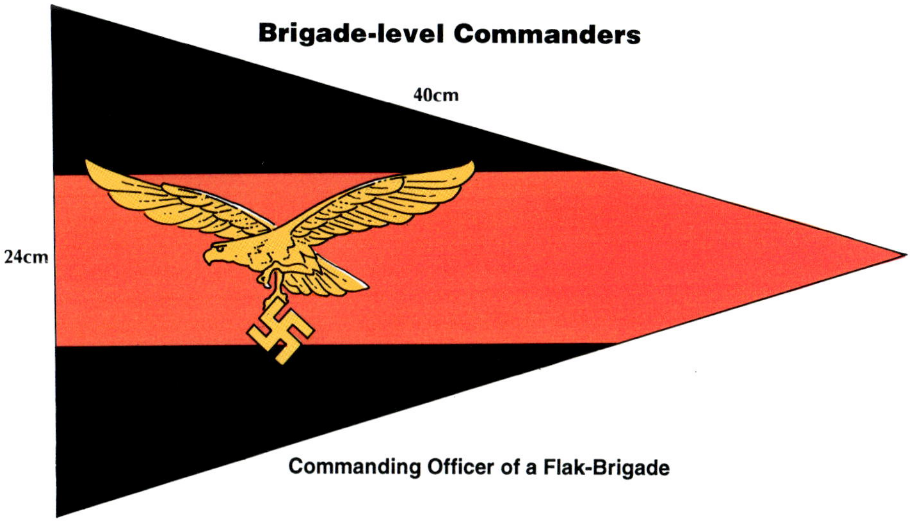

After 13.7.1940:

(a) Commanding Officer attached to the Army High-Command and also Fighter Unit leaders (yellow)

(b) Commanding Officer of a Flak-Brigade (rcd)

(c) Signals Officer attached to a Luftgau Command with disciplinary powers of a brigade commander (brown)

Staff Flags

Regiment-Level Commanders

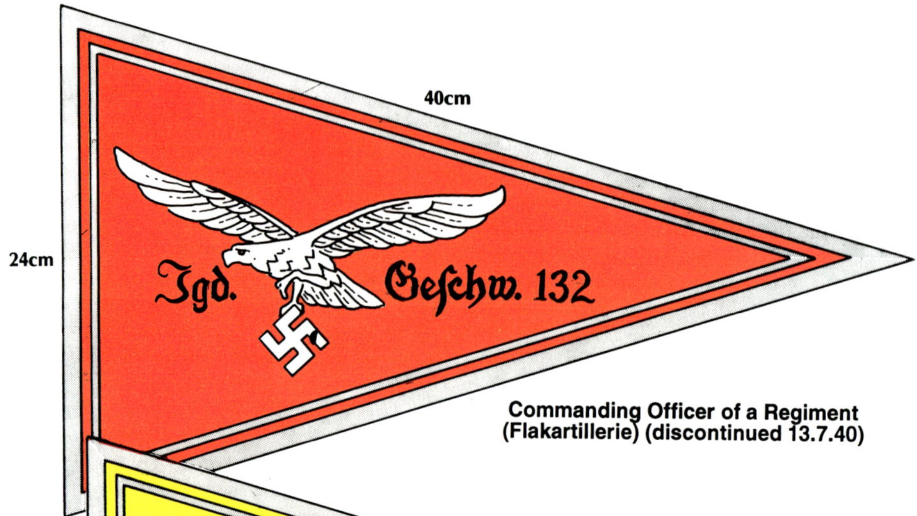

Commanding Officer of a Regiment (Flakartillerie) (discontinued 13.7.40)

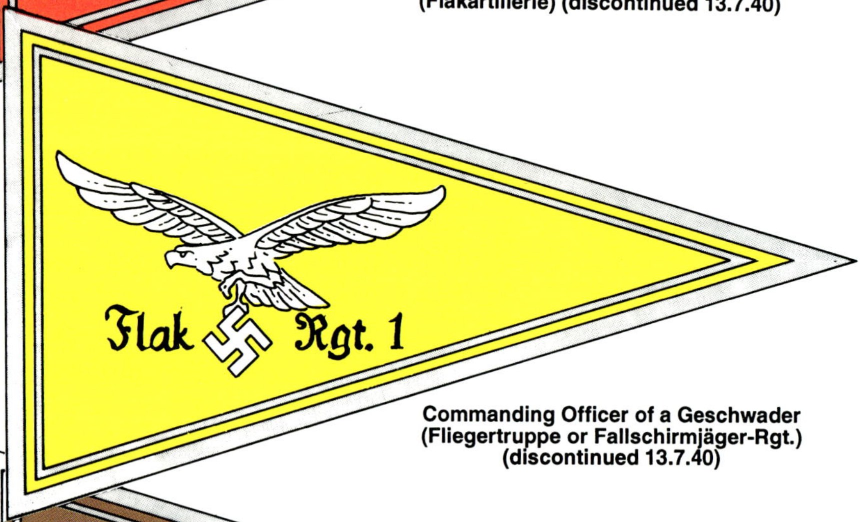

Commanding Officer of a Geschwader (Fliegertruppe or Fallschirmjäger-Rgt.) (discontinued 13.7.40)

Commanding Officer of a Regiment (Luftnachrichtentruppe) (discontinued 13.7.40)

After 13.7.1940:

(a) Commander of a Geschwader or Fallsch.Jg.Rgt. (yellow)
(b) Commander of a Flak Regiment (red)
(c) Commander of a Signals Regiment (brown)
(d) Commander of Rgt. "General Göring" (white)
(e) Signals Officer attached to a Luftgau Command with disciplinary powers of a regimental commander (brown)
(f) Signals Officer attached to a Flying-Corps or a Flak-Corps (brown)

Additionally after 16.7.1941:

(g) Transport Column Leader as commander of mixed units (yellow)

Commander of a Signals Regiment

Commander of a Supply Unit (after 29.4.43)

Additionally after 29.4.1943:

(g) Commander of a Light Infantry Regiment (rifle green)
(h) Commander of an Artillery Regiment (red)
(i) Commander of a Supply Unit (light blue)
(j) Commander of a Regular Infantry Regiment (white)
(k) Commander of a Guard Regiment (white)

Luftwaffe Field-Divs.

(l) Commander of a Light Infantry Regiment (rifle green)
(m) Commander of a Panzer Regiment (pink)
(n) Commander of an Artillery Regiment (red)
(o) Commander of a Flak Regiment (red)
(p) Commander of a Supply Unit (light blue)

Division "Hermann Göring"

Commander of a Panzer Regiment (Division "Hermann Göring")

NOTE:
1. The above Staff Flag was not for commanders of Flight Training Regiments, Schools, Demonstration Units and Flight Stand-By of the Luftwaffe High-Command.
2. Fallsch. Truppe had all flags in yellow Waffenfarbe.

Battalion-level Commanders

Discontinued 13.7.1940:

(a) Commander of a Flying Wing or Fallsch.Jg.Btl. (yellow)*

(b) Commander of a Flak Battalion (red)*

(c) Commander of a Signals Battalion (brown)*

(d) Commander of a Battalion in Regiment "General Göring"

- unit inscription for Rifle Battalion (green)
- unit inscription for Flak Battalion (red)
- unit inscription for Guard Battalion (yellow)

*unit inscriptions in black

Commander of a Flak Battalion (discontinued 13.7.40)

After 13.7.1940:

(a) Commander of a Flying Wing or Fallsch.Jg.Btl. (yellow)
(b) Commander of a Flak Battalion (red)
(c) Commander of a Reserve-Flak Battalion (red)
(d) Commander of a Signals Battalion (brown)
(e) Commander of a Battalion in Regiment "General Göring"

Additionally after 16.7.1941:

(f) Commander of a Supply Battalion or a Transport Column Battalion (yellow, red or brown)

(g) If the above is composed of mixed units (Fliegertruppe, Flakartillerie and Luftnachrichtentruppe), the branch color yellow is used.

Additionally after 29.4.1943:

(h) Commander of a Light Infantry Battalion (rifle green)
(i) Commander of an Artillery Battalion (red)
(j) Commander of a Flak Battalion (red)
(k) Commander of an Antitank Battalion (pink)
(l) Commander of a Signals Battalion (brown)
(m) Commander of a Supply Unit (light blue)

Luftwaffe Field-Divs.

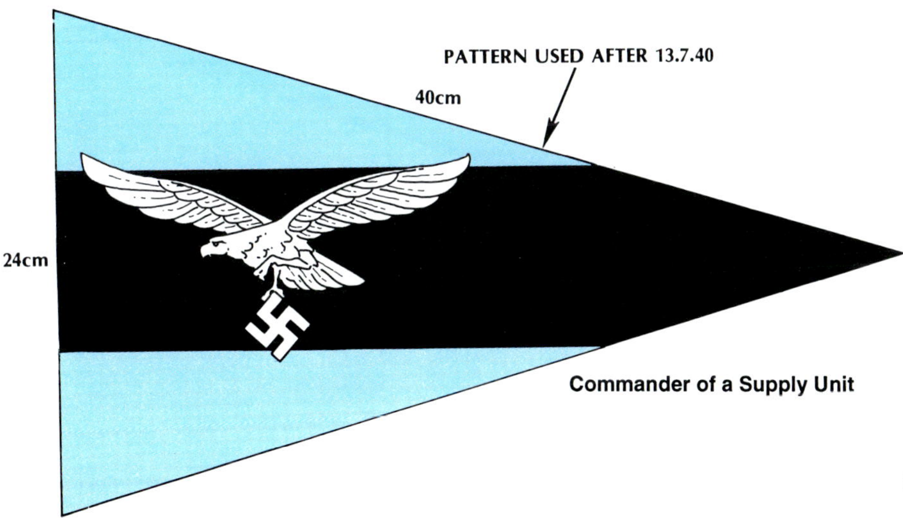

Commander of a Supply Unit

Commander of a Panzer Reconnaissance battalion (Div."H.G,") or Flying Wing, Fallsch.Jg.Btl., etc.

Division "Hermann Göring"

(n) Commander of a Regular Infantry Battalion (white)
(o) Commander of the Guard Battalion (white)
(p) Commander of a Light Infantry Battalion (rifle green)
(q) Commander of a Panzer Battalion (pink)
(r) Commander of a Panzer Reconnaissance Battalion (golden yellow)
(s) Commander of a Flight Stand-By (golden yellow)
(t) Commander of an Artillery Battalion (red)
(u) Commander of a Flak Battalion (red)
(v) Commander of the Führer Flak Battalion (red)
(w) Commander of a Signals Battalion (brown)
(x) Commander of a Supply Unit (light blue)

Commander of a Light Infantry Battalion

NOTE:

The above Staff Flag not for commanders of Flight-Training Battalions, Replacement Signals Battalions, Replacement Flak Battalions, Schools, Demonstration Units, Medical Battalions and Test Sites.

Company-Level Commanders

After 13.7.1940:

(a) Commanding Officer of an Independent Squadron (yellow)

(b) Commanding Officer of an Independent Flak Battery (red)

(c) Commanding Officer of an Independent Signals Company (brown)

(d) Commanding Officer of an Independent Company from Regiment "General Göring" (white)

NOTE:
The above are independent units which did not belong to a Wing or Battalion.

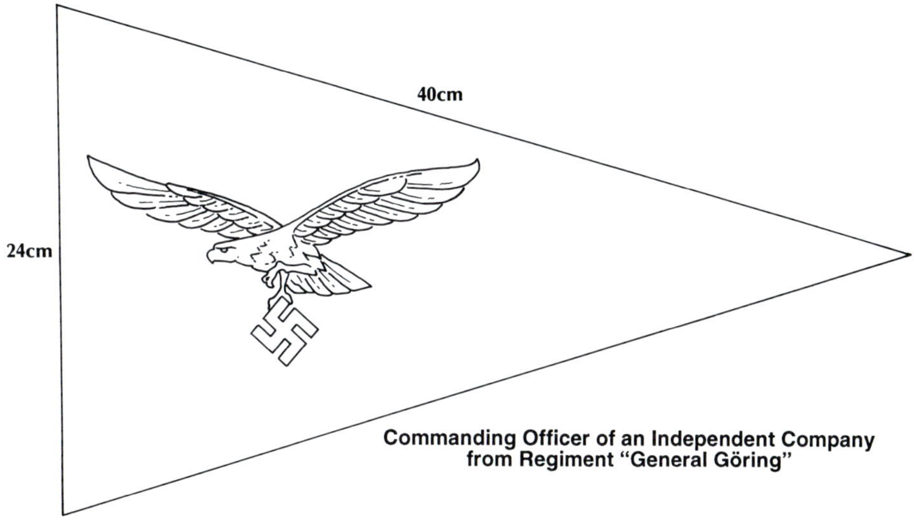

Commanding Officer of an Independent Company from Regiment "General Göring"

Commanding Officer of an Independent Flak Battery

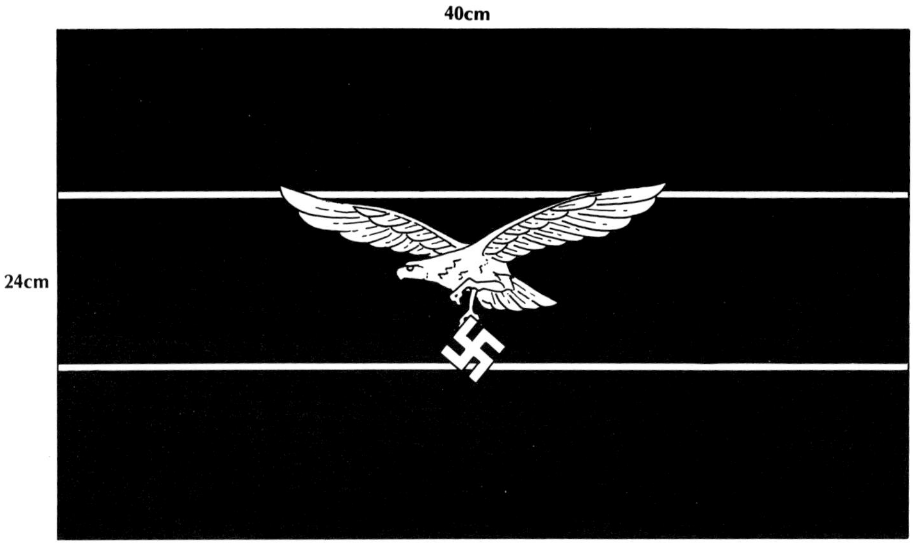

Commander of a Construction Engineer Regiment (after 2.4.43)

After 2.4.1943:

(a) Commander of a Luftwaffe Construction Engineer Battalion

Additional after 29.4.1943:

(b) Commander of an Engineer Battalion (Luftwaffe Field Division)
(c) Commander of an Engineer Battalion (Division "Hermann Göring")

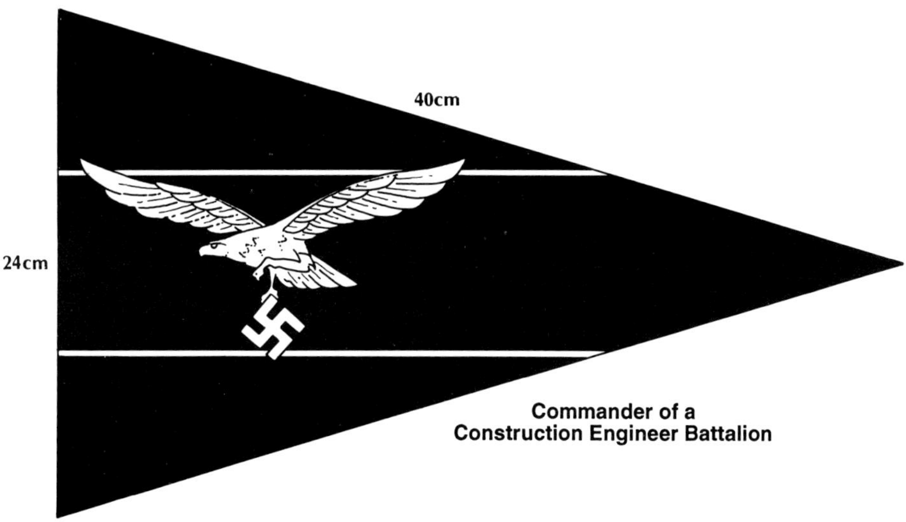

Commander of a Construction Engineer Battalion

Commander of an Air-Defense Brigade (after 6.11.43)

Commander of an Air-Defense Regimental staff (after 2.4.43)

Commander of an Air-Defense Battalion (motorized) (after 2.4.1943)

Car Pennants

It should be noted that the preceeding command and staff flags were utilized in car pennant form. When the officer was uniformed and in his official car, the command or staff flag (if authorized) was positioned on the left front fender. If the officer was not authorized a command or staff flag and was uniformed and in his official car, the Luftwaffe car pennant was positioned on the left front fender. When traveling in their personal vehicle, officers, officials and technicians of officer rank had the Lufwaffe car pennant positioned on the right front fender.

The Luftwaffe in the Sudetenland (note pennant on left fender).

General Officers and Officials of General Ranks (yellow eagle)

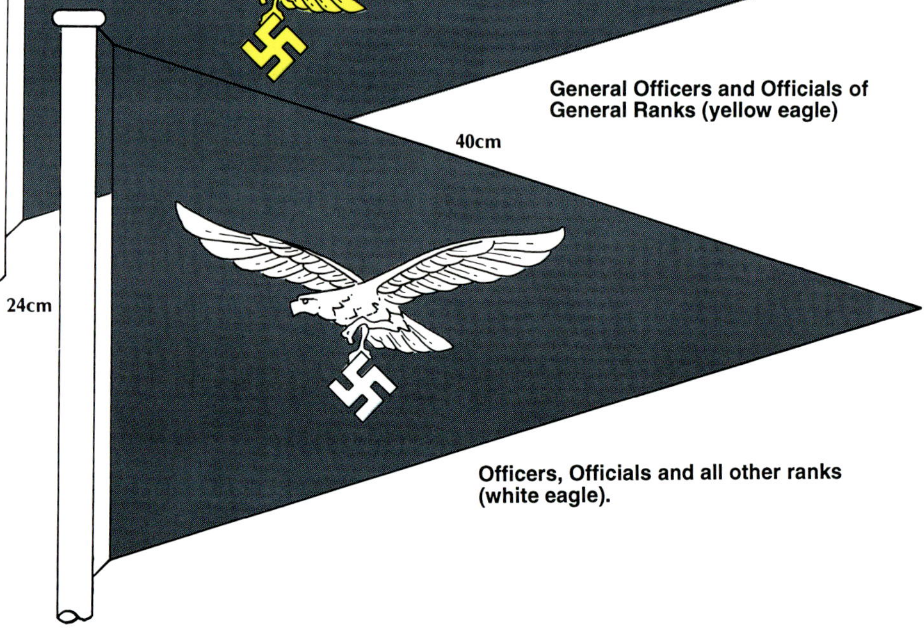

Officers, Officials and all other ranks (white eagle).

After 7.2.1941:

The following Luftwaffe car pennant was for all general officers (if not using a command or staff flag) including medical, officals (Wehrmachtbeamte) and technical officers (Ingenieurkorps). As of 7.2.1941 it replaced the pennant with yellow eagle (see preceeding page). All other officers and officials continued to use the grey-blue Luftwaffe pennant with white eagle.

After 10.9.1942:

The following Luftwaffe car flag was for all general officers in command positions and replaced the above pennant which was introduced in 1941. (It was not, however, for commanding generals of a Flying-Corps or a Flak-Corps...see page 290).

An ME-110 from SKG210 (schnelles Kampfgeschwader – Fast Bomber Geschwader)

SKG210 EMBLEM

Military Aircraft Markings

1935-2.7.36 (all fighters, bombers, etc.)

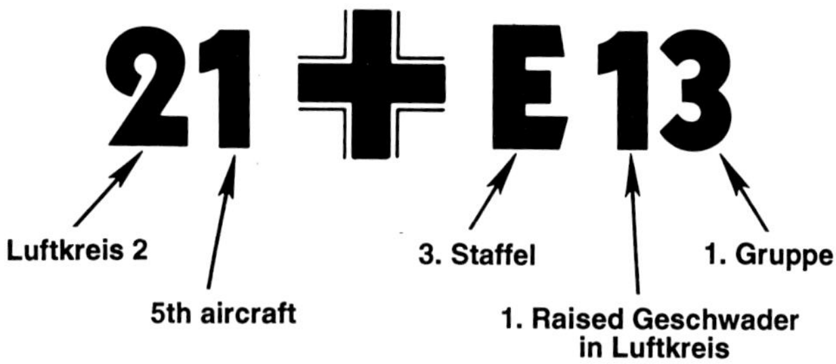

The first number of the code indicated the Luftkreiskommando (Local Air Command) or the Luftgaukommando (Provincial Air Command) which are listed below.

Luftkreis listing (from 1.3.35)
Luftkreiskommando 1 - Königsberg
Luftkreiskommando 2 - Berlin
Luftkreiskommando 3 - Dresden
Luftkreiskommando 4 - Münster
Luftkreiskommando 5 - München
Luftkreiskommando 6 - Kiel (sea)

Luftgau listing (from 5.39)
Luftgaukommando I - Königsberg
Luftgaukommando III - Berlin
Luftgaukommando IV - Dresden
Luftgaukommando VI - Münster
Luftgaukommando VII - München
Luftgaukommando VIII - Breslau
Luftgaukommando XI - Hannover
Luftgaukommando XII - Wiesbaden
Luftgaukommando XIII - Nürnberg
Luftgaukommando XVII - Wien

2.7.36-14.12.37 (fighters only)

The following Geschwader*/Gruppe**/Staffel*** markings were initiated on July 2, 1936 and were to be in effect by September 1 of that year (white outlined in black). Each Staffel aircraft carried an identification number 1 - 12 (also white outlined in black) which indicated the aircraft within a Staffel. The Geschwader staff, Gruppe staff and aircraft number markings were also carried on the upper surface of the top wing. The Staffel markings were carried on the aircraft as indicated below.

* Geschwader Markings:
The Geschwader were identified by traditional colors on the spinner, motor cowling and upper fuselage.

JG 131	- black (Jesau)
JG 132 "Richthofen"	- red (Döberitz)
JG 134 "Horst Wessel"	- brown (Dortmund)
JG 232	- green (Bernburg)
JG 233	- blue (Bad Aibling)
JG 234	- orange (Köln)

** Gruppe Markings:
I. Gruppe - no marking
II. Gruppe - horizontal bar
III. Gruppe - waved bar

*** Staffel Markings:

1st Staffel in a Gruppe - no marking

2nd Staffel in a Gruppe - white ring around the motor cowling and rear fuselage

3rd Staffel in a Gruppe - white discs on the motor cowling, fuselage sides and upper and undersurfaces

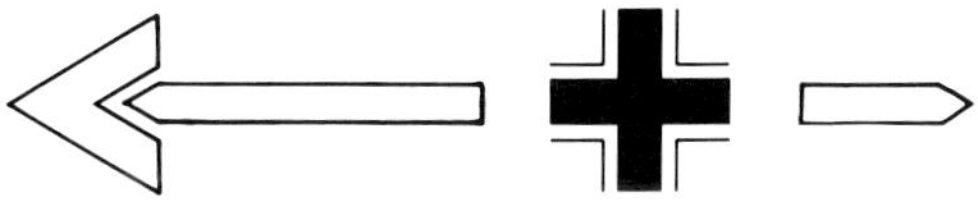

Gesch.Kdre.

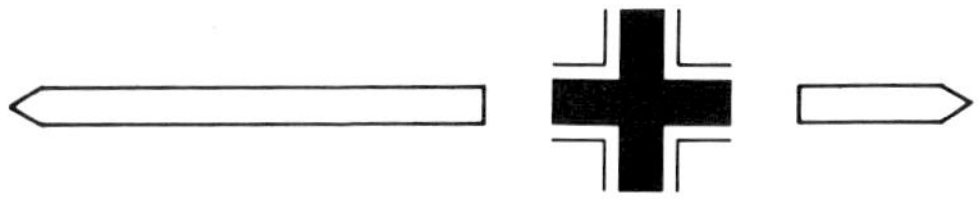

2nd plane, Gesch.Stab

3rd plane, Gesch.Stab

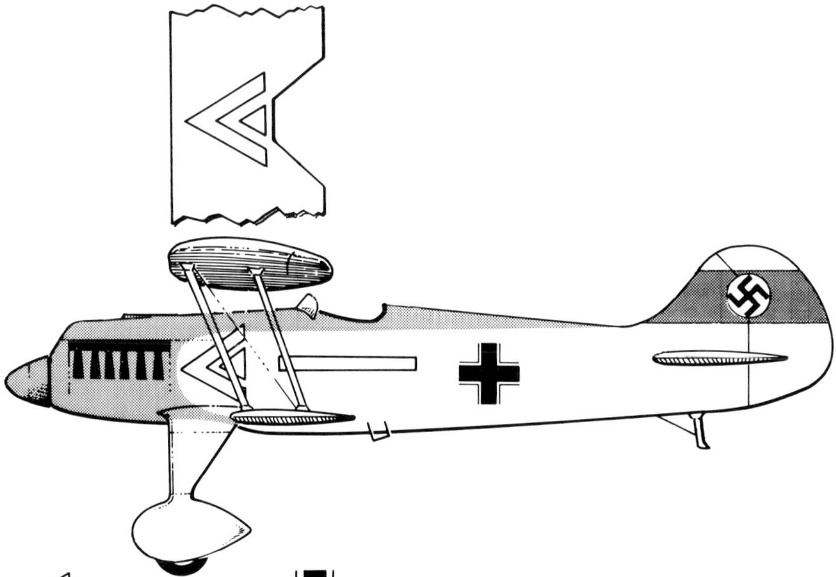

Gruppen-Kdr, II. Gruppe

Gruppe marking

2nd plane, Gruppenstab III. Gruppe

3rd plane, Gruppenstab I. Gruppe

The basic markings on German military aircraft were the Balkenkreuz on the fuselage sides and on the top and bottom of the wings plus a swastika on both sides of the rudder assembly.

Fighter Aircraft Markings

After 14.12.37, aircraft were marked with a number from 1 - 12 which designated the aircraft's number within a Staffel. The numerals were in the appropriate Staffel color (see below).

Staffel Identification (14.12.37)

(a) 1., 4. and 7. Staffel - numbers in white

(b) 2., 5. and 8. Staffel - numbers in red[1]

(c) 3., 6. and 9. Staffel - numbers in yellow[2]

in 1937, positioned in front of cockpit, 1938 and later in front of the Balkenkreuz

Gruppe Identification (14.12.37)

(a) I.Gruppe (Staffel 1 - 3) - no recognition symbol

(b) II.Gruppe (Staffel 4 - 6) - no horizontal bar

(c) III.Gruppe (Staffel 7 - 9) - vertical bar

(d) IV.Gruppe (Staffel 10 - 12)[3] - circle or small Balkenkreuz

emblem plus outline in Staffel colors

* Staff identification (Gruppe and Geschwader) in black

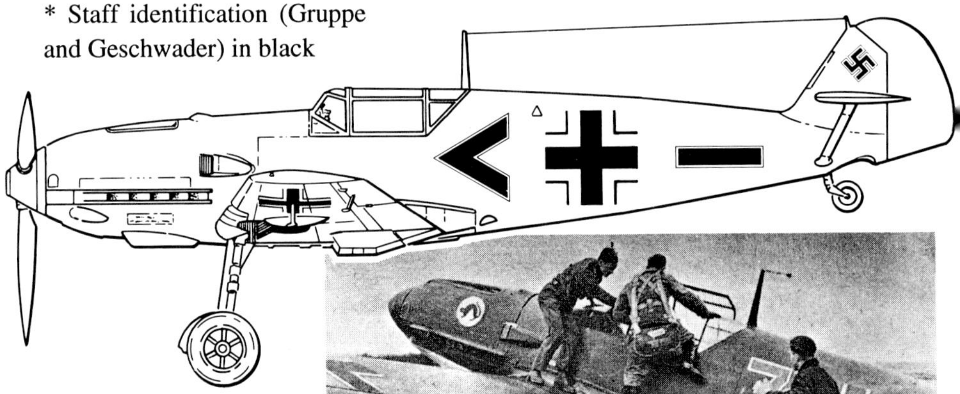

(above) Aircraft markings indicate flown by adjutant of the II.Gruppe (color of adjutant chevron and bar is black).

7th aircraft of the 8.Staffel, III.Gruppe JG 51 "Mölders" (indicated by cat insignia on cowling).

[1] Red numbers were outlined in white.

[2] Yellow numbers were outlined in black.

[3] The fourth Gruppe (Staffel 10 - 12) was introduced in 1940.

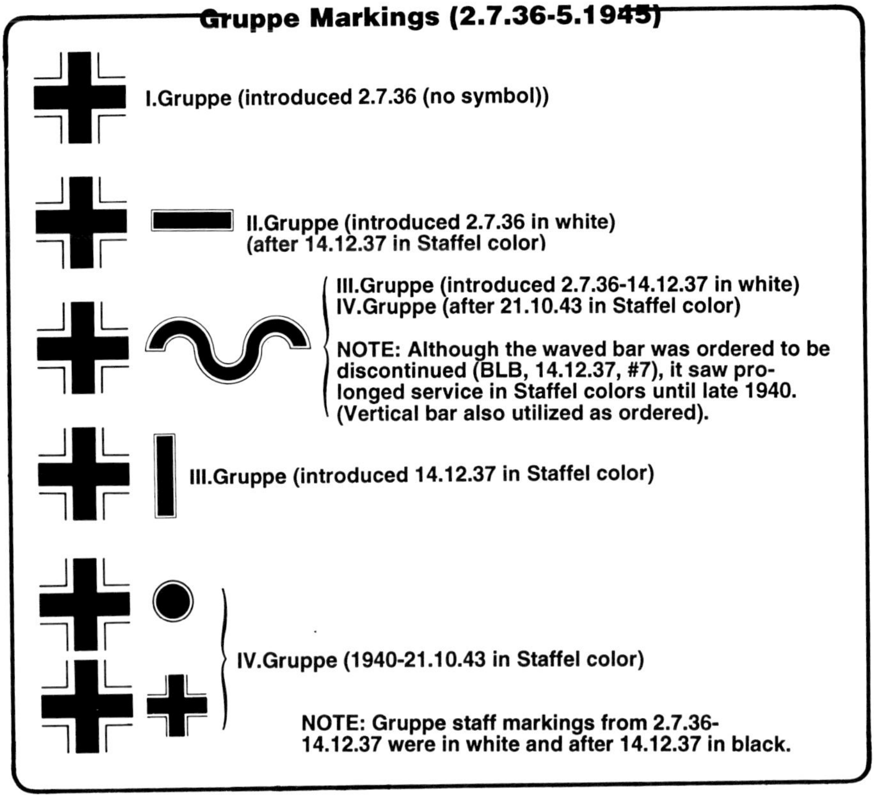

The 14.12.37 decree on aircraft identification markings was modified as follows on October 21, 1943. Staffel aircraft (including reserve aircraft) now carried the Arabic numbers 1 - 16 and were as follows:[4]

Staffel Identification

(a) 1. Staffel in the Gruppe - white numbers
(b) 2. Staffel in the Gruppe - black numbers
(c) 3. Staffel in the Gruppe - yellow numbers
(d) 4. Staffel in the Gruppe - blue with white outlined numbers

Gruppe Identification

(a) I.Gruppe - no Gruppe marking
(b) II.Gruppe - horizontal bar behind the Balkenkreuz in appropriate Staffel color
(c) III.Gruppe - vertical bar behind the Balkenkreuz in appropriate Staffel color
(d) IV.Gruppe - waved bar behind the Balkenkreuz in appropriate Staffel color with white outline

*Staff identification (Gruppe and Geschwader) in black

[4] Luftwaffen-Verordnungsblatt, November 8, 1943. The order is numbered 1988 and dated October 21, 1943.

(after 21.10.43)

*Gruppe staffs (black), otherwise in Staffel color

Staff Identification (14.12.37 and 21.10.43)

The Gruppe commander's aircraft carried in place of the arabic number, a black chevron with a smaller black triangle within it (outlined in white) as well as the Gruppe marking (I. - no marking, II. - horizontal bar, III. - vertical bar, IV. - waved bar) in black with a white outline. The other aircraft of the Gruppe staff conform to that of the Gruppe commander except they carried the chevron without the smaller black triangle.

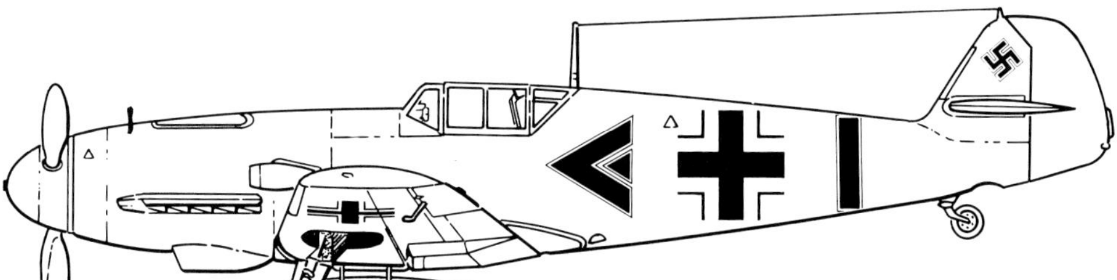

Gruppe commander's aircraft (III.Gruppe) (after 14.12.37)

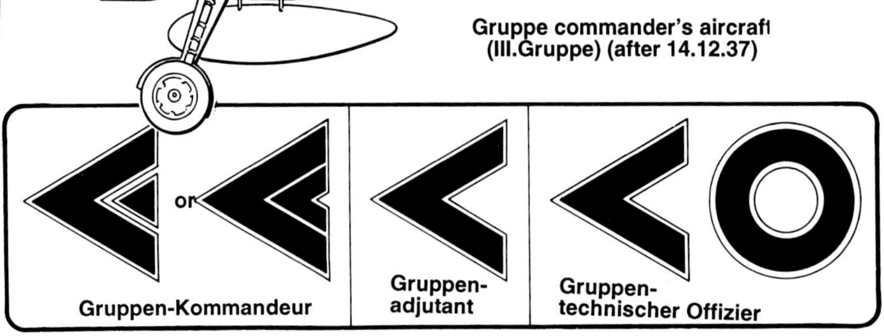

Geschwader Staff Identification (14.12.37 and 21.10.43)

The Geschwader commander and staff aircraft carried the markings illustrated below or variations thereof, until mid-1943. By an order dated April 14, 1943, all Geschwader staff markings were to be abolished. In their place a simplified form of marking was initiated. The Geschwader commander's aircraft carried the black chevron or arrow head and a pointed black bar outlined in white, in front of the Balkenkreuz. All other Geschwader staff aircraft carried a pointed, 970 mm long, black bar in front of the Balkenkreuz. Behind the Balkenkreuz, all staff aircraft carried a 1150 mm black bar outlined in white (see illustrations on next page).

Geschwader-Kommodore (variations)

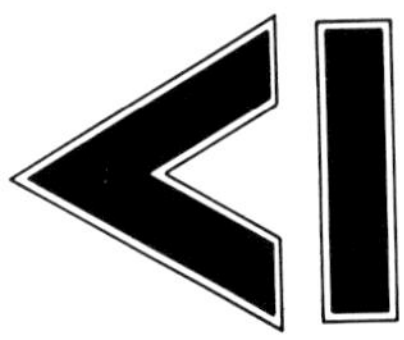

Geschwader-Adjutant

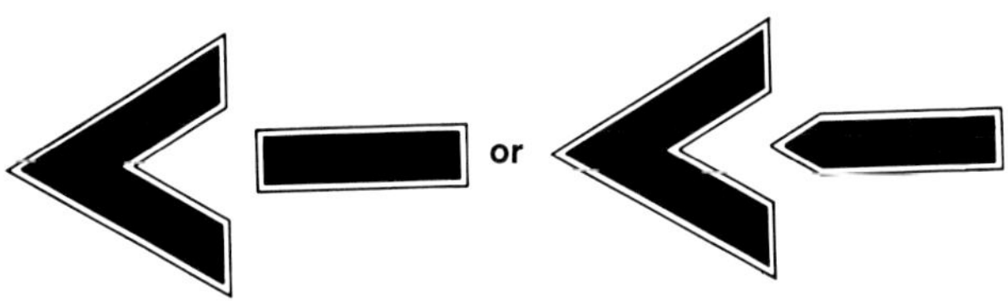

Geschwader Ia

Geschwader-Technischer Offizier

Major Beim Stab

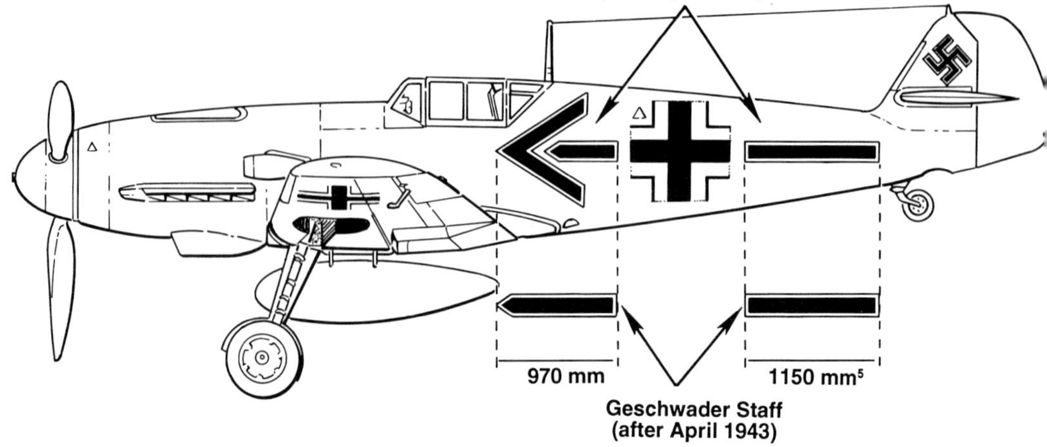

It should be noted that the regulations for Geschwader staff markings were not rigidly adhered to. Early and late staff markings (above and on the previous page) were sometimes combined or given a personal flair by individual pilots.

Propeller Spinner Markings

Aircraft spinners were generally colored in their respective Staffel color or in a combination Staffel/Gruppe color (also green for staff aircraft) as added identification. A few of the more frequent designs having been encountered are illustrated below:

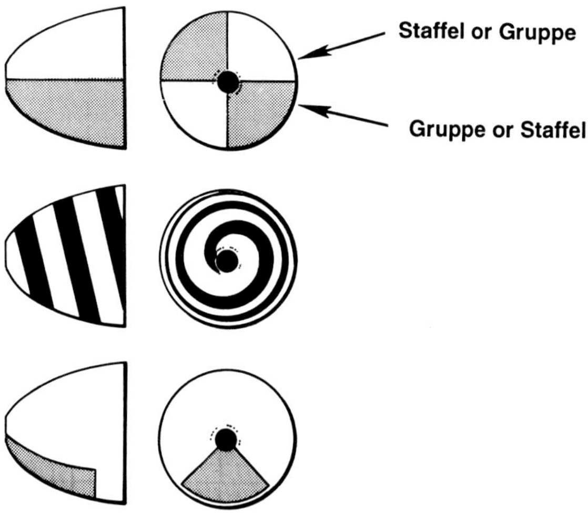

[5] Luftwaffen-Verordnungsblatt, April 27, 1943. The above specifications are from order number 788 and dated April 14, 1943.

The aircraft above are from the 1.Staffel (white numbers), I.Gruppe (no Gruppe markings behind Balkenkreuz) of JG27 (emblem on cowling). White fuselage ring indicates North Africa as the theater of operations.

JG 27 Emblem

Fuselage Rings

Colored rings around the fuselage, in front of the rudder assembly, indicated the theatre of operations. They were as follows:

from late 1941:[6]

white - North Africa, the Mediterranean area and southern Russia

yellow - central and northern Russia, the Scandinavian countries, the Channel area and the defense of the Reich

from early 1944:[7]

red/red parallel rings - defense of the Reich (fighters)

from January 1945 - the end of the war:

blue/white/blue parallel rings - defense of the Reich (fighters)

black/white/black parallel rings - defense of the Reich (variation)

[6] Ries Jr., Karl, vol.I, pp.13-14.

[7] Ries Jr., vol.II, p.5.

Ground Support and Fast Bomber Aircraft Markings

In the autumn of 1940, one Staffel from each Gruppe was refitted for fighter-bomber service. Soon thereafter, these Staffeln were formed into Schnelle Kampfgeschwader (fast bomber Geschwaders) and marked with a white falling bomb behind the Balkenkreuz in place of the Gruppe symbol.

By an order dated April 14, 1943,[8] the Staffel and Gruppe markings for ground support and fast bomber Geschwaders were modified. The Balkenkreuz and swastika were as before. Staffel aircraft, however, now carried the Latin letters A to Z and were as follows:

(a) 1., 5. and 9. Staffel - letters in white
(b) 2., 6. and 10. Staffel - letters in black
(c) 3., 7. and 11. Staffel - letters in yellow
(d) 4., 8. and 12. Staffel - letters in blue bordered in white

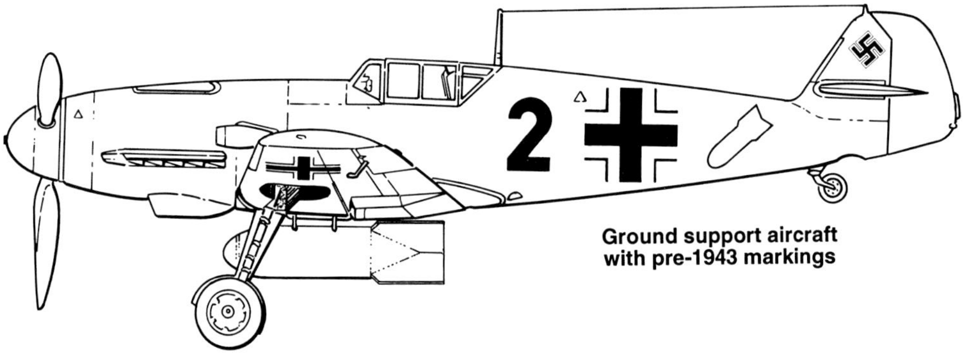

Ground support aircraft with pre-1943 markings

Gruppe identification

(a) I.Gruppe (Staffel 1 - 4) - no recognition symbol

(b) II.Gruppe (Staffel 5 - 8) - a horizontal bar behind the Balkenkreuz in appropriate Staffel color

(c) III.Gruppe (Staffel 9 - 12) - a vertical bar behind the Balkenkreuz in appropriate Staffel color

Gruppe Staff Identification

A Gruppe commander's aircraft carried in place of a Latin letter before the Balkenkreuz, a black chevron with a black triangle within it (both outlined in white). The Gruppe markings (I. - no marking, II. - horizontal bar, III. - vertical bar) in this case were also in black with a white outline. The other aircraft of the Gruppe staff were marked in accordance with the Gruppe commander, but without the black triangle within the chevron. (See Gruppe staff markings for fighter aircraft)

Geschwader Staff Identification

The identification markings for Geschwader commanders and their staff are as those described for fighter aircraft.

Long Range Fighter Aircraft Markings

In late 1938, long-range fighter units were formed to provide long-range protection. Initially Me 109 Cs and Ds were utilized for this purpose and were marked as fighter aircraft of that period. With the change-over to the Bf 110, the letter identification system of bomber formations was adopted (see bomber aircraft markings). This system was retained with the change-over to the Me 210 and later to the Me 410. The known Geschwader codes (letter/number combination) before the Balkenkreuz are as follows:

M8 = ZG 76
U8 = ZG 26 "Horst Wessel" (after July 1941 - 3U)
3M = ZG 2 (later 2S)
6U = ZG 1 (later 2J)

Night Fighter Aircraft Markings

The first night fighter Geschwader (NJG 1) was organized in 1940 and utilized in the defense of the Reich and in long-range night fighting. The aircraft involved, Bf 110s and Ju 88s, were painted black until mid-1941. From that time they were camouflaged in light and dark grey. With this camouflage system, the Balkenkreuz and swastika were either in a light or dark grey or completely painted over. Apparently the only universal emblem used by all NJGs employed as night fighters was the night fighter Geschwader emblem. The following Geschwader codes were sporadically employed by night fighter formations.

C9 = NJG 5
D5 = NJG 3
G9 = NJG 1
R4 = NJG 2
W7 = NJG 100
3C = NJG 4 (from January - July 1943, NJG 6)
7J = NJG 102
9W = NJG 101
2Z = NJG 6 (from January - July 1943, 3C)

The Night Fighter Geschwader emblem (above right) consisted of a white eagle on a red lightning bolt striking England.

Bomber Aircraft Markings

In late 1938, an identification system was introduced for bomber Geschwader and was used until the end of the war. It consisted of a letter/number combination in front of the Balkenkreuz and two letters behind it. The letter/number combination in front of the Balkenkreuz represented the Geschwader. The first letter behind the Balkenkreuz, usually in the Staffel color or black and outlined in the Staffel color, indicated the aircraft within the Staffel. This letter was repeated in black on the undersurface of the wings, near the tips.[9] The last letter of the rear combination indicated the Staffel or Staff within a Geschwader. The known Geschwader codes (letter/number combination before the Balkenkreuz) are:

A1 = KG 53 "Legion Condor"
A3 = KG 200
B3 = KG 54 "Totenkopf-Geschwader"
F1 = KG 76
F8 = KG 40
G1 = KG 55 "Greifen-Geschwader"
U5 = KG 2 "Holzhammer-Geschwader"
V4 = KG 1 "Hindenburg"
1G = KG 27 "Boelcke"
1H = KG 26 "Löwen-Geschwader"
1T = KG 28
3Z = KG 77 (formerly KG 153)
5J = KG 4 "General Wever"
5K = KG 3 "Blitz-Geschwader"
6N = KG 100
9K = KG 51 "Edelweiss-Geschwader"

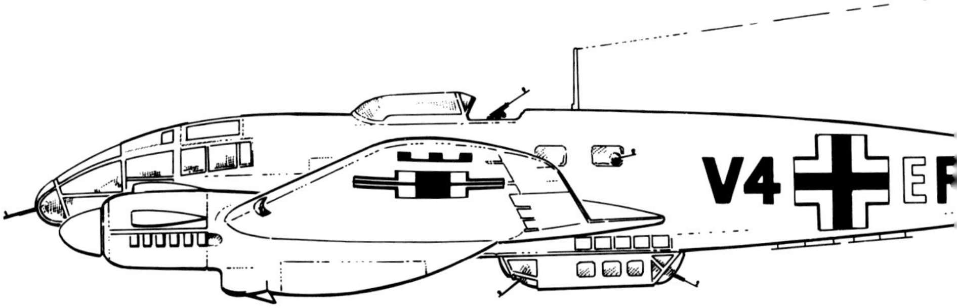

HE III H from the 7.Staffel, III Gruppe, Kampfgeschwader "Hindenburg"

[9] The first letter behind the Balkenkreuz was in blue for Geschwader staff aircraft and in green for Gruppe staff aircraft. The blue/green staff coding was utilized also by aircraft of long-range fighter, dive bomber, ground support and transport Geschwader. Karl Ries Jr., *Markierungen und Tarnanstriche der Luftwaffe im 2. Weltkrieg* (vol. II), (Finthen bei Mainz: Verlag Dieter Hoffman, 1965)., p.5. It must be pointed out that this series is the finest work of its kind and has been of great assistance in research as well as confirming my own findings.

"Löwen" Geschwader Crest

The 1H number/letter combination indicates the "Löwen" Geschwader.

Geschwader and Gruppe Staff Identification (last letter)

- A = Geschwader staff (blue)
- B = I.Gruppe staff (green)
- C = II.Gruppe staff (green)
- D = III.Gruppe staff (green)
- E = IV.Gruppe staff (green)
- F = V.Gruppe staff (green)

Gruppe and Staffel Identification (last letter)

I.Gruppe (white
- H = 1. Staffel (white)
- K = 2. Staffel (red)
- L = 3. Staffel (yellow)

II.Gruppe (red)
- M = 4. Staffel (white)
- N = 5. Staffel (red)
- P = 6. Staffel (yellow)

III.Gruppe (yellow)
- R = 7. Staffel (white)
- S = 8. Staffel (red)
- T = 9. Staffel (yellow)

IV.Gruppe (blue)
- U = 10. Staffel (white)
- V = 11. Staffel (red)
- W = 12. Staffel (yellow)

V.Gruppe
- X = 13. Staffel (white)
- Y = 14. Staffel (red)
- Z = 15. Staffel (yellow)

NOTES:

1. The Staffel colors above were for the first letter behind the Balkenkreuz.
2. After 1943, the Geschwader codes were painted very small in front of the Balkenkreuz or completely deleted. The code also appeared in a very small form on the rudder assembly.

Stuka and Ground Support Aircraft Markings

Stuka Geschwader adopted the letter/number code system which was used by all flying formations with the exception of the fighter units. Known Stuka letter/number codes are as follows:

H7 - StG 3
J9 - StG 5
S2 - StG 77
T6 - StG 2 "Immelmann"
6G - StG 51

Ground support units carried not only the badge of the unit but an infantry assault insignia in white under the cockpit or on the nose of the aircraft. From 1942 to mid-1944, ground support aircraft (Hs 123, Hs 129 and Fw 199) bore a solid black triangle outlined in its appropriate Staffel color in front or in back of the Balkenkreuz.[10] The Ju 87 did not carry this emblem. The Gruppe identification system for fighters was adopted in the spring of 1944, replacing the black triangle. A code number or letter in the appropriate Staffel color was positioned in front of the Balkenkreuz. This practice was continued for the duration of the war.

A HS123 with the white infantry assault badge on its cowling.

A HS129 from the II.Gruppe of the 1.Schlachtgeschwader

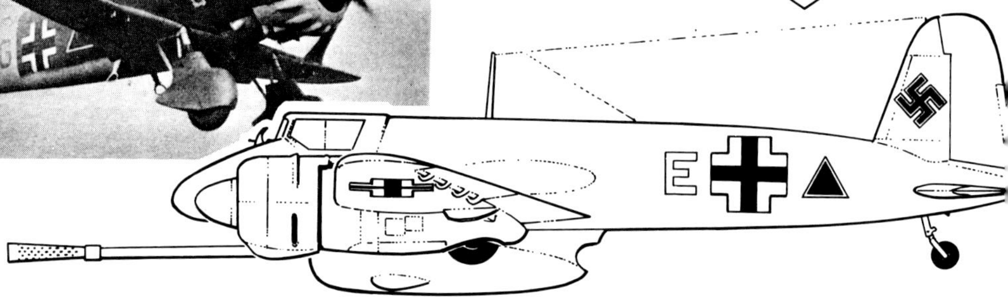

[10] There were two seperate units of the ground support wings (Schlachtgeschwader). The first wing consisted of a Geschwader staff and two Gruppen. If the black triangle was positioned in front of the Balkenkreuz, the aircraft belonged to I.Gruppe and if it was behind the Balkenkreuz, the aircraft was a part of the II.Gruppe. The individual aircraft letter was in its appropriate Staffel color which was as follows:

I.Gruppe	II.Gruppe
1. Staffel - white	5. Staffel - white
2. Staffel - red	6. Staffel - red
3. Staffel - yellow	7. Staffel - yellow
4. Staffel - blue	8. Staffel - blue

The second wing (with the exception of Ju 87 equipped units) utilized the Staffel and Gruppe markings described on page 312 under "Ground Support and Fast Bomber Aircraft Markings," after April 1943.

Reconnaissance Aircraft Markings

Reconnaissance Gruppen adopted the Geschwader letter/number code and Staffel letter system (as for bomber formations) at the outbreak of hostilities in 1939. Since reconnaissance Gruppen often consisted of six Staffeln, the Staffel colors (white, red, yellow, and blue) were not sufficient and therefore more often than not the Staffel colors were omitted from the first letter after the Balkenkreuz. Some of the known Reconnaissance Gruppe codes are as follows:

A6 - Aufkl.Gr. 120
F6 - Aufkl.Gr. 122
G2 - Aufkl.Gr. 124
P2 - Aufkl.Gr. 21
T1 - Aufkl.Gr. 10 "Tannenberg"
V7 - Aufkl.Gr. 32
7A - Aufkl.Gr. 121
5D - Aufkl.Gr.31
5F - Aufkl.Gr. 14
8H - Aufkl.Gr. 33
5M - Aufkl.Gr. 122
4N - Aufkl.Gr. 22
4U - Aufkl.Gr. 123

Transport Aircraft Markings

The Ju 52/3m served two purposes in pre-war Germany, either as a bomber or with slight modifications as a transport for troops and supplies. While on transport duty they were, therefore, still considered bombers, but on special assignment...hence the abbreviation KGzbV (Kampfgeschwader zur besondern Verwendung). After 1940, this designation was changed to TG (Transportgeschwader). The first letter after the Balkenkreuz was painted in the Staffel color as well as the leading edges of the wings outside both motor mountings. The Staffeln were identified by the letters A to Z as were bombers. Some known letter/number codes are as follows:

C8 - TG 5
G6 - KGzbV 2
L5 - KGzbV 5
1Z - KGzbV 1 (later TG 1)

Additional Aircraft Marking Information

An order dated December 15, 1938 and effective January 1, 1939, stated that the red/white/black national flag emblem which had been on both sides of military aircraft rudder assemblies were to be replaced by the swastika only. It was to be black with a white border (the border to be 1/6 the width of the swastika). The white border was also bordered with a thin black line, 1/4 the width of the white border.[11]

[11] Luftwaffen-Verordnungsblatt, January 30, 1939. Order #25.

Bibliography

Angolia, John R., DAGGERS, BAYONETS AND FIGHTING KNIVES OF HITLER'S GERMANY, Mountain View: R. James Bender Publishing, 1971.

ANZUGORDNUNG FÜR DIE LUFTWAFFE, Berlin: Verlag "Offene Worte," 1935 and 1935 (revised 1.4.37).

Denzel, Egon, DIE LUFTWAFFEN-FELDDIVISIONEN 1942-1945, 1963.

DER ADLER, (the official Luftwaffe magazine), 1939-1944.

DEUTSCHER LUFTWAFFENKALENDER 1941, Bamberg: J.M. Reindl, 1941.

Eickelbaum, Dr., DAS BUCH VON DER LUFTWAFFE, Berlin: Verlaghaus Bong, 1940.

HANDBOOK ON GERMAN MILITARY FORCES (TM-E 30-451), Washington: United States Government Printing Office, 1945.

Hettler, Eberhard, UNIFORMEN DER DEUTSCHEN WEHRMACHT, Berlin: Verlag Otto Dietrich, 1938.

Hettler, Eberhard, NACHTRAG 1939/1940 - UNIFORMEN DER DEUTSCHEN WEHRMACHT, Berlin: Verlag Otto Dietrich, 1940.

JAHRBUCH DER DEUTSCHEN LUFTWAFFE, 1936-1941. Leipzig: Breitkopf & Härtel.

JOINT ARMY AND NAVY PUBLICATION (J.A.N. 1), Assistant Chief of Staff, G-2, and Director of Naval Intelligence.

Littlejohn, David and Dodkins, Col.C.M., ORDERS, DECORATIONS, MEDALS AND BADGES OF THE THIRD REICH (including the Free City of Danzig), Mountain View: R. James Bender Publishing, 1968.

LUFTWAFFEN-FIBEL DES DEUTSCHEN JUNGEN, München: J.F.Lehmann's Verlag, 1943.

LUFTWAFFEN-VERORDNUNGSBLATT, 1935-1945.

Reibert, Dr. jur. W., DER DIENST-UNTERRICHT IM HEERE, Berlin: E.S. Mittler und Sohn, 1935.

Ries Jr., Karl, DORA-KURFÜRST UND ROTE 13 (vols. I-IV), Finthen bei Mainz: Verlag Dieter Hoffmann, 1964-1969.

Ries Jr., Karl, MARKIERUNGEN UND TARNANSTRICHE DER LUFTWAFFE IM 2. WELTKRIEG (vols. 1-3), Finthen bei Mainz: Verlag Dieter Hoffmann, 1963-1967.

Tschoeltsch, Major, DER DIENSTUNTERRICHT IN DER LUFTWAFFE, Berlin: E.S. Mittler und Sohn, 1936 and 1941.

Von Westarp, E.-J. Graf, TASCHENKALENDER FÜR DIE LUFTWAFFE, Grimmen in Pommern: Verlag Alfred Waberg, 1942.

....and countless others consulted but not cited.